Praise for *The Whale Warriors*

"Heller's eye-opening book . . . is both a riveting account of Heller's two months aboard the small, dilapidated trawler with a ragtag group of volunteers risking their lives to incapacitate a six-boat fleet of Japanese whalers and an explanation of the politics that keep commercial whalers operating."

—*The Malibu Times*

"Venerated adventure author Heller's account of the time he spent aboard the *Farley Mowat* as the vessel and its crew hunted the Japanese whaling fleet near Antarctica in 2005. . . . Heller's writing captures the very real danger of tangling with massive vessels, determined whalers, and the weather at the bottom of the world as he builds a tension-laden sea tale complete with a gallery of salt-sprayed characters. . . . If you're looking for a read that's two parts high-seas swashbuckle and one part inconvenient truth, this is it."

—*Surfer* magazine

"Heller frequently reports from the rough edges of the world, and *The Whale Warriors* takes him to the environmental equivalent of a war zone. . . .The book is a swift kick to any remaining complacency about the plight of our oceans."

—*National Geographic Adventure*

"Heller paints a passionate picture of the plight of the world's oceans and the creatures who dwell within them. The book almost certainly will raise the reader's consciousness and ire."

—*Rocky Mountain News* (Denver)

"[Heller] does a masterful job of balancing the journalistic details of this voyage with background—sympathetic, but not fawning—on Watson and his crewmembers and the larger issues that Watson's crusade raises."

—*Riverfront Times* (St. Louis)

"The adventure and the all-star cast of characters aside, the heart of this book is Heller's gripping account of the world's oceans. Aboard the *Farley Mowat,* Heller gains insight into the claim that if current fishing practices and pollution trends continue, 'every fishery in the world's oceans will collapse by 2048.'"

—*Sacramento News & Review*

"Heller's writing is energetic and bold, at times a swashbuckling adventure, at others a portrait of a determined eco-warrior, at others a heart-rending expose on the cruelty of whalers."

—*Publishers Weekly*

"A convincing, passionate account that both educates and infuriates."

—*Kirkus Reviews*

"An adventure more gripping than any novel."

—*AudioFile*

"Peter Heller has written a funny, angry, explosive book, which is as much high adventure at sea as it is a portrait of our relationship to the world's oceans. You're reminded of Ed Abbey's explosive lyrical prose, the antics of Robin Hood, and the wry eye of John Steinbeck. If you've ever wondered about life aboard a 'vegan attack vessel,' *The Whale Warriors* is your ticket. Heller's world here is so unusual, so wild, that you'd think he'd discovered it across the far-flung seas, and you'd be right."

Doug Stanton, author of *In Harm's Way*

"There are few human beings worthy of being recognized as heroes. Captain Paul Watson and his crews are in the van, and Peter Heller gives them their well earned due. Read the book and cheer—and weep!"

—Farley Mowat

"A gripping account . . . I found myself holding onto the desk as he recounted 40-degree rolls. . . . I have hundreds of whale books in my library, but this title easily earns a place in the top 10."

—*The Globe and Mail*

"Sometimes funny, almost always adrenaline-fuelled, and often reads like a 19th-century high-seas adventure."

—*Maclean's*

"Eloquent, riveting and more than a little disconcerting. . . . A high-seas adventure tale, complete with Force 8 gales, high-speed chases, clashes with authorities, villains, fuel and food shortages and an anti-hero compelling and complex enough to carry the fast-paced narrative to its ambiguous conclusion."

—*Canadian Geographic*

Also by Peter Heller

Hell or High Water:
Surviving Tibet's Tsangpo River

Set Free in China:
Sojourns on the Edge

THE
WHALE
WARRIORS

*The Battle at the Bottom of the World to
Save the Planet's Largest Mammals*

PETER HELLER

Free Press
New York   London   Toronto   Sydney   New Delhi

Free Press
An Imprint of Simon & Schuster, Inc.
1230 Avenue of the Americas
New York, NY 10020

This Free Press trade paperback edition December 2017

Free Press and colophon are trademarks of Simon & Schuster, Inc.

For information about special discounts for bulk purchases,
please contact Simon & Schuster Special Sales at
1-866-506-1949 or business@simonandschuster.com.

The Simon & Schuster Speakers Bureau can bring authors to your live
event. For more information or to book an event, contact the Simon &
Schuster Speakers Bureau at 1-866-248-3049 or visit our website at
www.simonspeakers.com.

Map © 2007 by Jeffrey L. Ward

Book design by Ellen R. Sasahara

Manufactured in the United States of America

3 5 7 9 10 8 6 4

The Library of Congress has cataloged the hardcover edition as follows:

Heller, Peter.
The whale warriors : the battle at the bottom of the world to save the
planet's largest mammals / Peter Heller.
p. cm.
1. Whaling—Antarctic Ocean. 2. Whales—Conservation—Antarctic
Ocean. 3. Heller, Peter—Travel—Antarctic Ocean. 4. Heller, Peter—
Diaries. 5. Farley Mowat (Trawler). I. Title.
SH382.5.H45 2007
333.95'95—dc22   2007006983

ISBN 978-1-4165-3246-0
ISBN 978-1-5011-9376-7 (pbk)
ISBN 978-1-4165-4613-9 (ebook)

All photos courtesy of Peter Heller unless otherwise noted.

To Darling Kim

To Caila, Zoë, Camryn
who are not at all meek
and who shall inherit the earth anyway

To Jackson, beloved

Contents

Lone-flier screams
Urges onto the whale-road
The unresisting heart across the waves of the sea.

   —The Seafarer

He makes flat warre with God, and doth defie
With his poore clod of earth the spacious skie.

   —George Herbert, "The Church Porch"

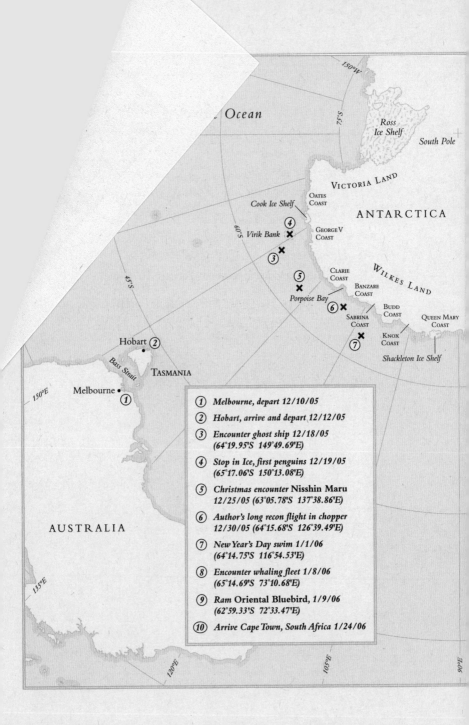

Ocean

150°W

75°S

Ross
Ice Shelf

South Pole

VICTORIA LAND

OATES
COAST

Cook Ice Shelf

ANTARCTICA

60°S

Virik Bank

④ ✕

GEORGE V
COAST

③ ✕

WILKES LAND

45°S

CLARIE
COAST

⑤

BANZARE
COAST

BUDD
COAST

✕

QUEEN MARY
COAST

Porpoise Bay

⑥ ✕

SABRINA
COAST

✕

KNOX
COAST

⑦ ✕

Shackleton Ice Shelf

Hobart ②

Bass Strait

TASMANIA

150°E

Melbourne •

①

① *Melbourne, depart 12/10/05*

② *Hobart, arrive and depart 12/12/05*

③ *Encounter ghost ship 12/18/05*
(64°19.95'S 149°49.69'E)

④ *Stop in Ice, first penguins 12/19/05*
(65°17.06'S 150°13.08'E)

⑤ *Christmas encounter Nisshin Maru*
12/25/05 (63°05.78'S 137°38.86'E)

⑥ *Author's long recon flight in chopper*
12/30/05 (64°15.68'S 126°39.49'E)

⑦ *New Year's Day swim 1/1/06*
(64°14.75'S 116°54.53'E)

⑧ *Encounter whaling fleet 1/8/06*
(65°14.69'S 73°10.68'E)

⑨ *Ram Oriental Bluebird, 1/9/06*
(62°59.33'S 72°33.47'E)

⑩ *Arrive Cape Town, South Africa 1/24/06*

AUSTRALIA

135°E

120°E

105°E

90°E

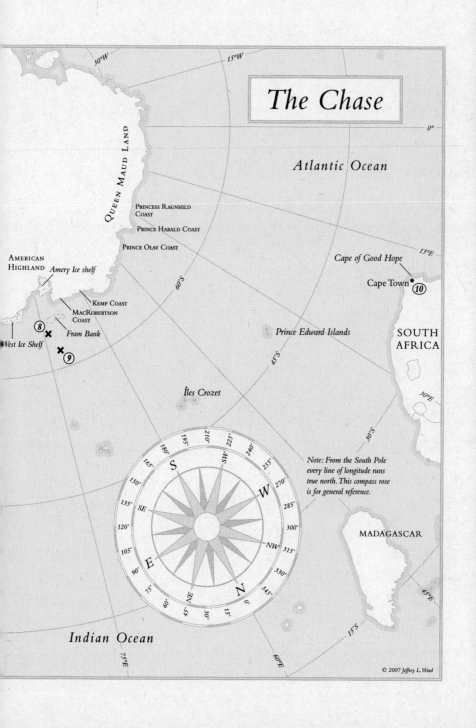

1

Storm

At three o'clock on Christmas morning the bow of the *Farley Mowat* plunged off a steep wave and smashed into the trough. I woke with a jolt. The hull shuddered like a living animal and when the next roller lifted the stern I could hear the prop pitching out of water, beating air with a juddering moan that shivered the ribs of the 180-foot converted North Sea trawler.

We were 200 miles off the Adélie Coast, Antarctica in a force 8 gale. The storm had been building since the morning before. I lay in the dark and breathed. Something was different. I listened to the deep throb of the diesel engine two decks below and the turbulent sloshing against my bolted porthole and felt a quickening in the ship.

Fifteen days before, we had left Melbourne, Australia, and headed due south. The *Farley Mowat* was the flagship of the radical environmental group, the Sea Shepherd Conservation Society. The mission of her captain, Paul Watson, and his forty-three member all-volunteer crew was to hunt down and stop the Japanese whaling fleet, which was engaged in what he considered illegal commercial whaling. He had said before the trip, "We will nonviolently intervene," but from what I could see of the preparations being conducted over the last week, he was readying for a full-scale attack.

I dressed quickly, grabbed a dry suit and a life jacket, and ran up three lurching flights of narrow stairs to the bridge. Dawn. Or what passed for it in the Never-Night of antarctic summer: a murky gloom of wind-tortured fog and blowing snow and spray—white eruptions that tore off the tops of the waves and streamed their shoulders in long streaks of foam. When I had gone to sleep four hours earlier, the swells were twenty feet high and building. Now monsters over thirty feet rolled under the stern and pitched the bow wildly into a feature-less sky. The timberwork of the bridge groaned and creaked. The wind battered the thick windows and ripped past the superstructure with a buffeted keening.

Watson, fifty-five, with thick, nearly white hair and beard, wide cheek bones, and packing extra weight under his exposure suit, sat in the high captain's chair on the starboard side of the bridge, look-ing alternately at a radar screen over his head and at the sea. He has a gentle, watchful demeanor. Like a polar bear. Alex Cornelissen, thirty-seven, his Dutch first officer, was in the center at the helm, steering NNW and trying to run with the waves. Cornelissen looked too thin to go anyplace cold, and his hair was buzzed to a near stubble.

"Good timing," he said to me with the tightening of his mouth that was his smile. "Two ships on the radar. The closest is under two-mile range. If they're icebergs they're doing six knots."

"Probably the *Nisshin Maru* and the *Esperanza*," Watson said. "They're riding out the storm." He was talking about the 8,000-ton Japanese factory ship that butchered and packed the whales, and Greenpeace's flagship, which had sailed with its companion vessel the *Arctic Sunrise* from Cape Town over a month earlier, and had been shadowing and harassing the Japanese for days. Where the five other boats of the whaling fleet had scattered in the storm no one could say.

Watson had found, in hundreds of thousands of square miles of Southern Ocean, his prey. It was against all odds. Watson turned to Cornelissen. "Wake all hands," he said.

■ ■ ■

In 1986 the International Whaling Commission (IWC), a group of seventy-seven nations that makes regulations and recommendations on whaling around the world, enacted a moratorium on open-sea commercial whaling in response to the fast-declining numbers of earth's largest mammals. The Japanese, who have been aggressive whalers since the food shortages following World War II, immediately exploited a loophole that allows signatories to kill a certain number of whales annually for scientific research. In 2005, Japan, the only nation other than Norway and Iceland with an active whaling fleet, decided to double its "research" kill from the previous year and allot itself a quota of 935 minke whales and ten endangered fin whales. In the 2007/2008 season it planned to kill fifty fins and fifty endangered humpbacks. Its weapon is a relatively new and superefficient fleet comprising the 427-foot factory ship *Nisshin Maru*; two spotter vessels; and three fast killer, or harpoon, boats, similar in size to the *Farley Mowat*.

Lethal research, the Japanese say, is the only way to accurately measure whale population, health, and its response to global warming and is essential for the sustainable management of the world's cetacean stocks. The director general of Japan's Institute of Cetacean Research (ICR), Hiroshi Hatanaka, writes, "The legal basis [for whaling] is very clear; the environmental basis is even clearer: The marine resources in the Southern Ocean must be utilized in a sustainable manner in order to protect and conserve them for future generations." Though the ICR is a registered nonprofit organization and claims no commercial benefit from its whaling, critics scoff, pointing out that the meat resulting from this heavily subsidized research ends up in Tokyo's famous Tsukiji fish market, and on the tables at fancy restaurants. By some estimates, one fin whale can bring in $1 million.

Each year the IWC's Scientific Committee votes on whaling proposals, and at its annual meeting in 2005 it "strongly urged" Japanese whalers to obtain their scientific data "using nonlethal means," and expressed strong concern over the taking of endangered fins, and humpbacks from vulnerable breeding stocks. The whalers' response was silence, then business as usual.

Although this resolution is not legally binding, much of the public was outraged that the whalers would openly disregard it. The World Wildlife Fund contended that all the research could be conducted more efficiently with techniques that do not kill whales. New Zealand's minister of conservation, Chris Carter, among others, described the Japanese research as blatant commercial whaling. Even dissenters within Japan protested: Mizuki Takana of Greenpeace Japan pointed to a report issued in 2002 by the influential newspaper *Asahi* in which only 4 percent of the Japanese surveyed said they regularly eat whale meat; 53 percent of the population had not consumed it since childhood. "It is simply not true that whaling is important to the Japanese public," Takana said. "The whaling fleet should not leave for the antarctic whale sanctuary."

To Watson there is no debate. The Japanese whalers are acting commercially under the auspices of "bogus research" and therefore are in violation of the 1986 moratorium. Even more controversially, the whaling occurs in the Southern Ocean Whale Sanctuary, an internationally ordained preserve that covers the waters surrounding Antarctica as far north as 40°S and protects eleven of the planet's thirteen species of great whales. Although research is permitted in the sanctuary, commercial whaling is explicitly forbidden. The whalers are also in clear conflict with the Convention on the International Trade of Endangered Species (CITES). And although the killing area in 2006 lay almost entirely within the Australian Antarctic Territory, the Australians, while protesting, seemed to lack the political will to face down a powerful trading partner. It irks Watson that Australian frigates will eagerly pursue Patagonian toothfish poachers from South America in these same waters, but will turn a blind eye to the Japanese whalers. "It sends a message that if you're rich and powerful you can break the law. If the Australian navy were doing its job," he said, "we wouldn't be down here."

Watson has no such diplomatic compunctions. He said, "Our intention is to stop the criminal whaling. We are not a protest organization. We are here to enforce international conservation law. We don't wave banners. We intervene."

Whaling fleets around the world know he means business. Watson has sunk eight whaling ships. He has rammed numerous illegal fishing vessels on the high seas. By 1980 he had single-handedly shut down pirate whaling in the North Atlantic by sinking the notorious pirate whaler *Sierra* in Portugal and three of Norway's whaling fleet at dockside. He shut down the *Astrid* in the Canary Islands. He sank two of Iceland's whalers in Reykjavik harbor, and half the ships of the Spanish whaling fleet—the *Isba I* and *Isba II*. His operatives blew open their hulls with limpet mines. To his critics he points out that he has never hurt anyone, and that he has never been convicted of a felony in any country.

2

Prelude

I had first met Watson the May before, at the Telluride Mountain Film Festival. He stood in front of 1,000 people in a large auditorium and told a story.

"In June 1975, sixty-five miles off the coast of Siberia, Bob Hunter and I ran our Zodiac between a Russian whaler and a small pod of panicked, fleeing gray whales. We were the first to use a Zodiac in this way. The whalers fired a harpoon over our heads and hit a female whale in the head. She screamed. There was a fountain of blood. She made a sound like a woman's scream. Just then one of the largest males I've ever seen slapped his tail hard against the water and hurled himself right at the Soviet vessel. Just before he could strike, the whalers harpooned him too. He fell back and swam right at us. He reared out of the water. I thought, this is it, it's all over, he's going to slam down on the boat. But instead, he pulled back. I saw his muscles pull back. It was as if he knew we were trying to save them. As he slid back into the water, drowning in his own blood, I looked into his eye and I saw recognition. Empathy. What I saw in his eye as he looked at me would change my life forever. He saved my life and I would return the favor."

Silence. Watson let the image sink in like a rhetorical harpoon.

He then said that the Japanese were still aggressively whaling and that he was going to go after them in his ship. He said the ocean is dying: of seventeen global fishing hot spots like the Grand Banks, sixteen have collapsed beyond repair. He said there are now only 10 percent of the fish stocks that were in the ocean in 1950. He said we should stop, every one of us, eating all fish. There was an uncomfortable stirring in the crowd.

"I don't give a damn what you think of me," he thundered. "My clients are the whales and the fish and the seals. If you can find me one whale that disagrees with what we're doing, we might reconsider."

Watson had been among the founders and first board members of Greenpeace in 1972. His encounter with the gray whale off Siberia had been part of Greenpeace's first voyage to protect whales, and he had served as first officer. In 1977 he broke away to form the Sea Shepherd Conservation Society. The legend among Greenpeacers is that he was thrown out for advocating violence, and for physically separating a sealer from his club, but Watson claims he was voted off the board of directors because he opposed the board presidency of Patrick Moore. He says he started Sea Shepherd because he wanted to specialize in direct interventions against illegal exploitation of the ocean.

In the last thirty years, Sea Shepherd has been running almost continuous campaigns at sea to stop illegal whaling, drift-netting, longlining, dolphin slaughter, and sealing. The organization, which is based in Washington state, spends no money on fund-raising, but gets donations through media attention and word of mouth. Pierce Brosnan, Martin Sheen, and Christian Bale are generous supporters, as are John Paul DeJoria, CEO of the Paul Mitchell hair products company; Yvon Chouinard, founder of Patagonia; and Steve Wynn, Las Vegas hotel and casino operator. Watson quipped, "With James Bond, the president, and Batman on my side, how can I lose?"

At the end of his talk Watson invited me to come with him to Antarctica as a journalist, and I accepted.

3

The Farley Mowat

December 5, 2005

The Australian customs official at the Melbourne airport looked up for the first time since scanning my passport and paperwork.

"You're staying in Australia 'A couple of days—not sure'?"

"Right."

"What does that mean?"

"Well, several days. Probably."

"Local address the *Farley Mowat*. Never heard of it."

"It's a ship."

"What kind of a ship?"

"It belongs to the Sea Shepherd Conservation Society. It's a conservation ship."

A flicker of suspicion. "Like Greenpeace, is it?"

"No, sir, not exactly." It would not be prudent to tell this official that Sea Shepherd made Greenpeace look like Sunday school. Or that several sovereign nations, including Norway and Japan, as well as the U.S. National Fisheries Institute, had leveled charges of piracy against

Sea Shepherd and its officers. The Norwegian navy had depth-charged and badly damaged the society's last ship. The Soviet navy, when there was one, had once come within seconds of machine-gunning a Sea Shepherd crew into the Bering Sea. Only the miraculous appearance of a gray whale surfacing and blowing between the ships had defused the situation.

"What then, exactly?"

"Have you seen Jacques Cousteau?"

The man's eyes did not move. The tip of his tongue touched the corner of his mouth. He was wavering between getting mean and deciding to enjoy himself. He waited. It was a bad analogy anyway. Watson was about as far from the benign, grandfatherly Cousteau as you could get. Cousteau loved people. Watson once said that the life of a human being is worth less than the life of a worm. Time and again he had offered up his own life, and the lives of his crew, in defense of a whale. He had written, "The pyramids, the Old masters, the symphonies, sculpture, architecture, film, photography . . . All of these things are worthless to the Earth when compared with any one species of bird, or insect, or plant."

The officer waited. "I'm on assignment for National Geographic *Adventure*," I said. "I'm covering the Sea Shepherd campaign against the Japanese whaling fleet in Antarctica."

Truth has a certain resonance. The man's face softened. "That sounds pretty interesting." His right hand went to the stamp.

"Antarctica, eh?" he said, handing me back my passport. "Be careful."

The taxi turned onto the street that ran along the wharf just at the foot of New Melbourne. All around the harbor, skyscraper condos were going up. There, tied up to the dock in the bright sun, between piers covered with the umbrellas of bars and restaurants, was a black hulk of a ship with a Jolly Roger flapping lazily from her bow. The skull was inset with a circling whale and dolphin, and instead of crossed bones, there were a shepherd's staff and a trident.

The *Farley Mowat*, while only one-third the length of the factory ship she would be hunting—and one-tenth the tonnage—was still over half a football field long. She radiated both nobility and menace. She was completely black, stem to stern. The only color was a nod to PR—the yellow letters on the side that said "seashepherd.org." She was low-slung forward of the bridge superstructure, where the main deck held three fast Zodiacs—inflatable outboard-motor boats—and two Jet Skis in their cradles. From the main deck the bow swept up to a gracefully rounded bludgeon of black steel. The hull was "ice-reinforced," meaning strong enough to push through moderately thick ice, and ideal for ramming. Water cannons bristled off the bow and the helicopter deck, which was a steel second level added over the aft deck. The cannons were there to prevent boarding. When I arrived at the dock the ship was crawling with crew, all dressed in black T-shirts. Knots of tourists and visitors milled on the pier and stared.

The immediate impression of the *Farley* was a ship of war. Whatever intentions her owners had when she was built in Norway in the summer of 1956, whatever fishing fleet pedigree was still expressed in her stout lines, she had been transformed by use. She had been reshaped by the will of her present captain. The *Farley* was a dedicated fighter. Not like any regulation navy ship, though: she had the maverick, menacing air of a privateer. No wonder the Japanese Institute of Cetacean Research had announced as the whaling fleet left Shimonoseki harbor on November 8 that they were afraid of an attack by Sea Shepherd.

The *Farley* was incongruous amid the prosperous bustle of the piers, the shiny condos and cafés. Two pickups with pallets of sacked food and pickle buckets pulled up to the gangplank amidships. Another truck backed up and three construction workers in union shirts got out and waited to talk with a skinny crew member on the wharf who was cupping his hands around a cell phone at his ear. He seemed to be the one in charge, so I lugged my duffels over the pavement and waited to introduce myself.

As I stood in the drubbing sun and inhaled the dense, intoxicating smells of harbor—of salt, rot, tar, diesel—my eyes wandered over the tough ship that would be home for the next month. The long black arm of the port davit, or crane, swung over the main deck and was being hooked to a rope harness on the port Zodiac. A young woman with wild curly blond hair who wore tattered greasy khaki shorts yelled orders from a raised platform that held the davit controls. She looked about nineteen. At the wheel of the Zodiac stood a strong kid with black athletic glasses and long red dreadlocks under a tied bandana; when he hopped up onto the inflated tube to catch the hook, I noticed that he had a fir tree tattooed on each calf. Next to him, in the center Zodiac, getting a harness ready, was a guy in a military buzz cut, talking into a radio strapped to his shoulder. He was in full black SWAT gear: cargo pants, black boots, and a tight military utility vest that held radio, knife, flashlight, and a dozen pockets.

On the tall superstructure, beneath the varnished gloss of fresh black paint and bubbles of rust, was the faint outline of columns of small figures. They were stenciled skulls and crossbones, and beneath them were the names of ships I recognized: *Isba I* and *Isba II*, the *Sierra*. The *Senet* and *Morild*. The litany of ships rammed and sunk. I counted fifteen in neat cemetery rows. Like the kills on the fuselage of an ace; a pentimento of scalps.

"You the National Geographic guy?" I turned. Another crew member: he had a deep tan, broad shoulders, pronounced cheekbones, and the grave eyes of a wolf. He had grease smeared on his neck, and another military haircut—a "high and tight" popular with special forces.

"Yeah. Hi."

"Let me show you your cabin."

He hefted the duffels like they were two pillows. I followed him to the edge of the dock where he jumped across a two-foot gap and through an open gate in the bulwarks or rail of the main deck. He turned left, aft, through the main hatch in the base of the superstructure, which led into a long narrow hallway leading farther astern. He took a sharp right and dropped down a set of steep steps into the bowels of the ship beneath the main deck. Another narrow companionway lit by fluorescent ceiling lights. Immediate press of close heat and mold, cut with the astringent sweetness of diesel fumes. One, two cabin doorways on the right. He pushed through a red curtain and swung the duffels onto the bunk. Someone else's gear was already on it.

"Steve." He shook my hand with a crushing grip. "I'll have Geert move his stuff."

"You ex-military?"

"101st, Airborne."

"Where?"

"Korea mostly. We patrolled the DMZ."

"Was that tough?"

The wolf eyes studied me for a second. He wasn't used to being asked a bunch of questions. He hesitated.

"Yeah, we never really got warm. The mission of my division was: Die in Place. To delay and engage the North Koreans in case of attack. That's what the men called it: 'Die in Place.'"

Steve vanished. I thought of the samurai mantra before battle: *You are already dead*. Anybody who believed that would be a tough adversary.

A moment later a tall, gangly biker with a bushy dark beard and leather vest tumbled into the tiny cabin. On the official crew manifest Geert Vons was listed as "Ship's Artist." Smiling over his beard, he gathered up an armful of battered pack, one dirty blanket and pillow, and a sketch pad and tumbled out again. A few minutes later I saw him on the dock. He was a tattoo artist in Amsterdam. He worked above the Hells Angels bar, and drove a motorcycle all year, through the Dutch winter that occasionally froze the canals. He had completed a degree in Chinese and was now doing research with a Chinese scientist on the Baiji dolphins of the Yangtze River. He created illustrated children's books on marine wildlife. He seemed like an unlikely Hells Angel associate.

This was Geert's third campaign with Sea Shepherd. He had been on the last fruitless Antarctica campaign in the winter of 2002, when the *Farley* had patrolled the ice edge for a month and never laid eyes on a Japanese ship.

We watched a knot of Japanese tourists reading the public education placard at the foot of the gangplank and then, remarkably, drop some money into the blue plastic whale. An older crew member was leading a line of high school volleyball players in red jumpers and plaid skirts onto the ship, and I heard him say, "We are not a protest organization. We are empowered by the UN Charter for Nature to uphold international conservation law—huh? No, those are not guns, they are water cannons . . ."

One of the union guys called out to the skinny officer, "Where you want the welding rod, mate?" I noticed that his shirt said, "Electrical Trade Union. If You Don't Fight, You Lose."

I asked Geert about the cannons.

"They can't put too much pressure in them, because the pipes are still quite old. But for effect it's still quite good."

The pipes wouldn't hold the pressure because they were rusted out—because the ship was fifty years old and was being run by an organization on a shoestring.

"The hull, the bottom—is that rusty too?"

"Last March, on the seal campaign, the hull had a strong leak. The two bilge pumps couldn't catch up. It was off the ice, Newfoundland. She might have gone down. Alex dived down with a wood—like a carrot—and plugged it. They found the bottom was all scattered with rust spots, thin like paper. They patched everything in Jacksonville," he said benevolently, patting me on the back.

Geert went off to his day job illustrating the logbook up in the chart room.

The crew had dropped two of the Zodiacs into the harbor and used the knuckle boom—the central crane—to open the steel doors to the fish hold beneath the deck. The kid with the tree tattoos was tearing around the inner harbor in the smaller Zodiac, but the larger boat wouldn't start.

A tall, large-boned young woman with black hair down to her waist explained that they had flipped it going into Pitcairn.

Out of the fish hold they were now hoisting the oddest contraption. The thing looked like a Monty Python creation, two parts engine and fan blade, one part rubber dinghy. A hulking man with buzzed gray hair and heavy steel glasses limped around the opening yelling orders in a twangy tenor. He caught his breath.

"What is that?"

"FIB. Flying inflatable boat."

"It flies?"

The man smiled tolerantly. "It has wings. Thirty-six-foot span—hoh!" The blond pulled back on a lever and the FIB thing dangled. The man squinted at it. "I'm an ultralight pilot. They sent me down to Florida to get familiarized with it. I learned how difficult it is to water-taxi. The hardest part is getting it out of here and on the water." He talked in a rush and then stopped abruptly. He peered up at the machine. "Theoretically, I should be able to handle bigger seas, and the helicopter can handle a lot more wind. Have to land and take off on the water. When there's enough wind for me to take off from the deck it's too windy to fly. I'll be limited to ten knots. I'll go up as high as I can. The helicopter can go one direction and I'll go the other."

I could not imagine getting the Thing out of its cocoon, and winged, and launched in the rough seas of the Southern Ocean. Much less flying in those winds. There are winds off of Antarctica that sailors call "sudden busters," gale-force blows that scream out of nowhere, wreak havoc, and vanish just as fast. You would not want to be circling around in the FIB Thing looking for the whalers when one of those kicked up,

"Chris Price," the pilot said, and held out a ham of a hand. "You can ride with me. That little seat behind the pilot. I'll hold your reservation." He grinned as if he was doing me the biggest favor in the world.

"Thanks."

I turned to go aft and bumped into a short, broad-shouldered, clean-cut guy in his mid-thirties.

"Crazy dude, huh?" he said. "Personally, I wouldn't use that thing as a stepladder."

This was the other pilot, of the helicopter, Chris Aultman. Before the trip I had asked Watson how he hoped to avoid the washout of the 2002 campaign. How did he know he would even find the Japanese whaling fleet in all that ocean?

Watson had responded almost cryptically, "Well, we're working on getting a helicopter." Now, apparently, they had one, though I hadn't seen it anywhere.

The "research" area the Japanese planned to hunt in the 2005–2006 season spanned 35°E to 175°E, in an arcing swath from the coast of Antarctica, out to 60°S, some 300 miles off the ice. Approximately 1.4 million square miles. This was an area of ocean bigger than the states of Colorado, New Mexico, Arizona, Utah, Wyoming, Montana, Idaho, Nevada, California, and Washington combined. Imagine that you are in a car in Denver and your job is to find a convoy of a semi-truck and five pickups in all that area. Which way do you go? You have radar, but it is really effective only out to about twenty miles. Your jalopy goes only ten miles an hour and you have only fifty days' worth of fuel. And something else: there are no more cities or towns left in the entire region. No restaurants, no grocery stores, no garages, not even a place to replace a blown tire or tie-rod. There are half a dozen widely scattered outposts manned by a few disinterested scientists. One more thing: among the provisions you brought with you, there is no meat or cheese or eggs. The *Farley Mowat* was a vegan ship. You wouldn't find in her holds even a single preserved fish.

Chris Aultman seemed just as eager as Price to talk about his craft, a little Hughes 300. He was an instructor and commercial pilot out of John Wayne Airport in Orange County. He had seven years' experience in a helicopter, but he'd never flown off a moving deck before.

We would be picking up the bird in Hobart, Tasmania on the way south. He said, "Due to economic restraints we do not have a mechanic on board. The helicopter just went through an annual mechanical inspection. Everything suspect was replaced. Made as ready as can be. From that point on—if it breaks, the show's over."

Of all the jobs on the ship, he had, by far, the most dangerous. Piloting a temperamental whirlybird out of an airport with a hangar and a full maintenance staff is brave enough. Flying off a spray-lashed deck on a corkscrewing boat in the absolute middle of nowhere is another thing altogether. The *Farley* only went nine or ten knots. If a "contingency" did occur and Aultman had to ditch, say, eighty miles from the ship, it would take eight or nine hours for the ship to even reach his vicinity—if, that is, the crew knew where the hell that was. Aultman said that the chopper had "fixed utility floats," or pontoons, so the protocol in the event was to stay in the aircraft as long as possible.

Just then a kid stepped up—a beanpole, maybe twenty-one years old, with a black ponytail and a nearsighted squint behind little wire-rimmed glasses. He was Peter Hammarstedt, the second officer, a Swede. He said, "Chris, sorry to interrupt. We've got some solid rubber for you. It's three centimeters."

The kid left.

"Donated rubber. Unbelievable. That stuff is so expensive."

"Who donated it?"

"The union guys. A bunch of them were working on top of that apartment building on the east side, up there. They looked down and saw the ship and got curious. We gave them a tour and ever since they've got about every trade union in Melbourne supporting us. Unbelievable. They opened their showers up the street. Just dropped off $10,000 worth of welding rod."

He explained that the rubber was for the helipad. "Pitch is not a problem." He rocked his hand forward and back. "It's the roll that's the problem. Side to side. On the way down here they got forty-degree rolls, if you can imagine. I'm going to plant the helicopter on the deck—no messing around making it pretty." He slapped his hand down on the rail. "But it takes a second or two for the full weight to get on the skids. If the ship rolls at that moment it will slide. That's where the rubber will help." He chewed his lip.

"The Japanese fleet is trying hard not to be found. They have very good listening equipment that gives bearing and distance. They also have very sophisticated radar. I've heard people say they have over-the-horizon capability, which is satellite-assisted. They have the technology to make sure they are not found. When Sea Shepherd was down here last time trying to find them, they knew what area of ocean they were in but the Japanese stayed over the horizon."

If Aultman was forced down on one of his sorties, I wondered how the officers of the *Farley* would ever find him. "They'll know where I'm going to be," he said. "I'll call them every five minutes with my position. That keeps the search area narrow. If something happens. That's about a five-square-mile search grid to find me. I'll also have Epurb on board. If everything shits the bed I turn the Epurb on and that'll give a search area smaller than my communication grid." An Epurb is an emergency signal beacon. None of this seemed very comforting. The water temperature down there was thirty to thirty-two degrees.

"Paul knows their killing fields, so to speak. They are thousands of square miles. Which field they'll be in is anyone's guess. I don't know if Paul has an insider somewhere—he's been petitioning the Australian government to get help with satellite data, but they refuse. They want to help, but not in an official capacity. He's also put an online reward of $10,000 to anyone that can hack the position of the fleet. The fleet has to talk to a base back home and that data is somewhere."

"I can see a lot of ocean. We're planning on a four-hour mission right now. At about seventy nautical miles per hour that gives me 280 nautical miles at straight-line distance—out and back or draw a box. Haven't worked out a search grid. Obviously, the higher you go the farther you can see. If I can get up to 3,000 to 5,000 feet I should be able to see sixty nautical miles on either side of the helicopter. That gives me a 120-nautical-mile path I can see. That's a lot of ocean."

But in an area the size of the American West, it was a hairbreadth.

Chris was an ex-marine, a Southern California kid who liked to surf the breaks north of San Diego; he was clean-cut, and he worked every day in an olive-drab flight suit, and carried a clipboard and an ops manual. He had taken two months off work, left his wife in Long Beach, and volunteered to risk his life along with a bunch of vegans.

"You must love whales," I said.

"They shouldn't be killing them. It's wrong and they shouldn't be doing it."

There was plenty of action on deck. Two of the hands were taking turns jumping into a garbage pail barefoot and marching in step. Some good Samaritan had donated a heap of ripe grapes. "We are making ze wine for ze New Year party!" one announced with glee. Note to self: drink no homemade wine on the *Farley Mowat*. Other crew members had gotten the FIB Thing out and into the water and Price was tearing up and down the inner harbor, albeit without wings. A woman and a guy were trying to start the center Zodiac. Red Dreads was doing slow doughnuts in one of the purple Jet Skis. It died and he hand-paddled it back to the rope ladder.

"Our armada's not looking too good," he called up. I asked the big-boned woman what was wrong with the Jet Ski. She had large Roma eyes. "Wrecked it going into Henderson Island. Got caught in a wave."

She smiled and held out a hand tattooed with a leaf. "Inde," she said. "That's my forest name. My normal name is Julie."

She said that she and Red Dreads, whose forest name was Gedden, aka Jon, did a lot of forest actions with Earthfirst! Mostly they set up tree sits, to protect areas of forest slated for logging. Last winter she had sat in a Douglas fir for six weeks.

"On a branch?"

She laughed and her face lit up. "On a platform suspended from ropes. I lead a lot of all-women workshops, teaching how to make safe sits, how to maintain them."

"How do you get the first rope up there?"

"Bow and arrow."

4

The Good Captain

I saw Watson ambling down the dock carrying two shopping bags. He was wearing the same khaki belted jacket I had seen him wearing when I first met him. The same shaggy gray hair and beard.

He looked jovial, like a man about to embark on a great vacation. In fact, he was about to go do what he loved most in the world—set off on the high seas with a committed crew to intervene in what he deemed illegal exploitation of marine species. Coming into Melbourne two weeks before, the port control had radioed to ask Watson what class of vessel the *Farley Mowat* was. "Yacht," he'd replied. Vessel class is a serious matter: merchant ships incur much higher port costs, from pilot to dock fees, and—perhaps more important—they require officer and security certifications that would be impossible for a craft like the *Farley* to obtain. Later, as the pilot brought the ship into the inner harbor, port control said, "*That* is a pleasure craft?" "Why, sure," Watson said. "We enforce international conservation law. I take great pleasure in it." If a harbormaster gave him a hard time, he pointed out that he carried no cargo or paid passengers. "Tiger Woods's yacht is bigger than we are, and carries more crew," he would say.

He stopped to talk to a reporter for a couple of minutes, and then went up the gangplank without casting more than a glance at the hubbub on the deck. He'd learned long ago to leave the details to his crew. I followed him to the bridge. On the way, I bumped into the volleyball team. The players were finishing up their long tour. Half of them seemed ready to enlist.

"What is that?" one girl asked, pointing up to an oddly shaped fixture wrapped and tied in canvas like a deck gun.

"That," said the guide proudly, "is a catapult. We fill it with rotten garbage. It has about a 100-foot range." The girls looked concerned. "It's all organic," he added quickly.

On the bridge, Watson was leaning out one of the small sliding windows conversing with his first mate, the skinny guy on the dock. The bridge was all business: a narrow, wood-lined control room that stretched from one side of the superstructure to the other. A band of thick Plexiglas windows wrapped the cabin. In front of the high captain's chair on the starboard side was an inset revolving pane, which allowed visibility in a lashing storm. At the captain's left hand, side-mounted, was a chrome lever that controlled the ship's speed. Sophisticated electronics were arrayed all across the forward bulkhead under the windows: three radar systems, Global Positioning System (GPS) screens, three radios with mics mounted to the ceiling or deck head, gyrocompass with electronic autopilot, a handheld electric control on the end of a thick cord with thumb-button rudder controls. Against the aft bulkhead, or back wall, was a single traditional brass spoked wheel for manual control of the rudder should hydraulics fail. Also, a little pedestal and a curious shelf.

On the pedestal was a green bronze bust of Poseidon. Above it, under a glass bell jar, was a small carved wooden figure of a winged demon, maybe four inches high; he was horned and roaring, very weathered, but ferocious withal. A little "Tibet" sticker marked the wall above him. Beside the demon was a fat plastic Buddha with the word "Lulo" scrawled in magic marker on his base.

"That's our religious corner," Watson said. "The wooden figure is the Hayagriva, 'the compassionate aspect of Buddha's wrath.' Back in

1986 I had a Tibetan monk come on the ship; he said, 'I have been asked to give this to you.' I said 'What is it?' 'I don't know.' I lashed it on the mast. It was there for many years. In 1989 I met the Dalai Lama and he told me he had sent it. He said it was a Hayagriva. The Tibetan scholar Robert Thurman explained: he said you never want to hurt anybody but sometimes you've got to scare the hell out of them."

About the plastic Buddha he said, "Marc the welder put it there. Lulo means 'Little Pecker.' That's our little pecker God."

"And that?" I pointed to the left of the Hayagriva.

"That's just a rubber duck."

He nodded out the window. As we watched, another pickup and a car pulled up with more donated food and supplies. A skinny girl in a very short summer dress got out. She was barefoot. She lit a cigarette and began unloading big plastic bags of homemade granola and bushels of potatoes and turnips. Then a one-ton truck with eighteen drums of engine lube oil backed up to the ship.

"This has been a very good city in terms of support," he said. "The union guys have been amazing. You say you need steel and half an hour later there's plates everywhere." He turned. "A couple of DC motors to fix," he said, "and we can leave. One's for cleaning the oil, the other's for starting the engine. Trouble with this ship is the out-dated DC systems which no one knows anything about. Turns out the only people who know anything are in Melbourne."

He tossed his head to get a thick shock of gray hair out of his eyes. "Find your cabin OK? It's a bit Spartan around here." I thought my cabin was deluxe. It had a desk, a sink, a full-size bunk, a closet the size of a school locker, and a glow-in-the-dark poster of the southern constellations taped to the ceiling. I'd half expected something from *Master and Commander*—a hammock slung among a score of others in a crowded 'tween decks.

But if Watson was concerned about how I'd like my accommodations, I understood. According to him one of the only other magazine journalists who had ever signed up for a Sea Shepherd campaign was in the late 1980s. After being shown to his own private cabin the young man had come back on deck and accosted Watson. "This is shit," he said. "I can't sleep there." Watson had given him the nicest cabin aside from his own and the first mate's. "You'll have to. It's all we've got." "Well, I won't!" The writer looked around at the ship, which was probably in the same state of chaos as the *Farley* was today, and said, "This whole ship is shit." Watson told the writer that *he* was shit and ordered him off the boat.

Media relations were a preoccupation with Watson. He'd written a chapter on it in his book *Earthforce! An Earth Warrior's Guide to Strategy*. He recognized that the only way for an environmental movement to accomplish anything was to get the word out. Ramming one illegal Taiwanese drift netter wasn't going to save the 300,000 whales and dolphins killed in those nets worldwide every year; or halt the 7 million to 20 million tons of "bycatch" thrown overboard—animals obscenely killed and wasted—by long-liners and trawlers and other fishing boats. Nor would sinking eight whalers stop whaling. But the international attention focused on the issue by the dramatic action of a group like Watson's might bring international pressure to bear. "If you have an action and no one covers it, it didn't happen."

He'd used the strategy masterfully in stopping the slaughter of baby harp seals on the ice of Newfoundland in 1984. Watson had staged several confrontations with sealers that drew media attention. And then he'd brought Brigitte Bardot onto the ice for her famous picture with the baby seal. The starlet on the cold snow holding the defenseless fuzzy white pup, and the pup's huge, trusting, liquid black eyes had iced it. An outcry ensued and sealing was shut down for ten years. But it came back with a vengeance. Every year since 1994 the Canadians have been killing up to 350,000 young seals, and almost every year Watson is there on the ice with his ship to try to intervene. It frustrated Watson that most people knew nothing about it. For years, the press seemed no longer interested. The recent advocacy of Paul McCartney was just beginning to renew public interest.

In Watson's book, in the chapter on "Preparations" the section headed "Deception" read like a credo. It was basically Watson's MO.

All Confrontation is based on deception. This is called the strategy of tactical paradox.

When you are able to attack, you must seem unable.

When you are active, you should appear inactive.

When you are near, you should have the enemy believe you are far. And when far, near.

Bait the enemy.

Pretend to be disorganized, then strike.

If the enemy is secure, then be prepared.

If the enemy be of superior strength, then evade.

If your opponent has a weakness of temper, then strive to irritate.

Make a pretense of being weak and cultivate your opponent's arrogance.

If your opponent is at ease, then ensure that they are given no rest.

If the forces of your opponent are united, then seek to divide them.

Attack when the enemy is unprepared.

Appear when you are not expected.

The leader who wins makes careful plans.

This section was immediately followed by one called simply, "Preparation for Death." "The Earth Warrior, like all warriors, must be prepared for death. Live each day fully as if it were your last. . . . Hoka hey—Lakota meaning: 'It's a good day to die.'"

A little farther on were the lines: "The nature of a government attack can be determined by your own strategy. If you utilize covert or illegal tactics, you can expect any direct or indirect strategies from government forces including your own assassination."

All this seemed a bit grandiose. A romantic hodgepodge, ancient wisdom of the East meets Crazy Horse meets the paranoia of a Trotskyite cell. Watson, after all, was not marshaling an army to attack the Tokugawa shogunate or occupy the Russian White House. But then he had used many of these very tactics to pull off, for instance, the complete blockade of the Canadian sealing fleet at the harbor entrance to the port of St. John's in Newfoundland in 1983. He had faced down the Soviet Union's navy, and had made one of the very few known unauthorized incursions onto Soviet soil along the Siberian coast without being molested. That was serious business.

Was he being paranoid about government attack and assassination? Greenpeace, during its campaign against nuclear testing by France in the South Pacific in 1985, had been infiltrated by a sexy French agent who supplied intel to a team of French commandos. On July 10 they blew up the Greenpeace ship *Rainbow Warrior* in Auckland harbor, killing a Dutch photographer. Sea Shepherd had been infiltrated by a slew of informers including an FBI agent who worked as an engineer, and an agent of the Canadian Surveillance and Intelligence Service. Of the FBI informer Watson said, "He did a pretty good job. If the FBI wants to pay people to work in our engine room that's fine with me."

The volleyball guide stepped onto the bridge with a heavy woman in a business suit.

"Captain? This woman would like to give you something."

The businesswoman stepped up, unsure how close she should get. She looked as though she'd just found herself in a cage with a tiger. "I'm a Melbourne woman," she said. "I've heard that you quite like a tipple, so I've got this." She held out a bottle of Johnnie Walker Red and I could see her plump hand shaking. "And this flask. Can you use it?" Watson took the gifts and said thank you without registering any emotion. There was an awkward moment, and the woman said, "All the best for the trip," and fled. Watson was absolutely incapable of making small talk.

I asked Watson, with so much at stake for the Japanese, and with so many visitors coming and going on his ship, if he ever worried about the kind of sabotage that took place on the *Rainbow Warrior*.

Watson looked out the bridge windows at the deck hands trying to get the second Jet Ski started. They weren't having much luck. "Greenpeace is so big today because of the *Rainbow Warrior* sinking. That kind of thing would backfire on them."

"You said in one of your books that you thought you'd meet your end at the hands of an assassin. Do you still think you're going to die by the sword?"

"Oh, I don't know. It doesn't matter anyway. Better than dying of some debilitating disease." He swiveled around. "You know why Doc Holliday was such a great gunfighter?" I shook my head. "Because he had tuberculosis. He left Chicago and went west to die as a gunfighter. He always had that edge. That was why he was better. The other guys always had that moment of hesitation." Watson chuckled. "And then he died in bed of tuberculosis."

Doc Holliday didn't go into a fight, though, with forty-three other people on his shoulders.

"Captain," I said, "how do you plan on finding the Japanese fleet?"

He glanced at me sideways and shifted gears smoothly. "No guarantees, of course, but I think we have a very good chance. I'll show you." He swung his feet onto the deck and led me through the doorway into the chart room just aft of the bridge. On the port side was a chart table, atop a cabinet of wide, heavy wooden drawers with brass handles that held charts of all the world's oceans. A small-scale chart of the whole of Antarctica and the Southern Ocean lay spread on the table, the white continent a ragged disk of ice with a tail of peninsula gesturing toward Cape Horn. A pencil and a parallel ruler lay on the chart.

"There are certain sectors they're operating in. I'm pretty confident this time. They've got to be 100 to 120 miles off the ice pack. That's where the whales are. This time of year the krill blooms are along the edge of the ice pack, and the whales are after the krill."

"The minkes, fins, and humpbacks are all baleen whales?"

"Yeah, but we call them piked whales; we don't call them minke whales. We try to get around naming whales after their killer. No matter what person you name them after, it's an insult. The right whale was named the 'right whale to kill.' Cachelot meant catch a lot."

He leaned over the chart.

"Here is Area 3. You're really looking at a corridor here. Bring the ship down the corridor and then you have aerial surveillance." He swept two fingers in an arc along the coast between Commonwealth Bay and the Shackleton Ice Shelf. "Greenpeace is heading from here, south of South Africa, where they're supposed to be. They're going to areas 2 and 3. That's where the Japanese are supposed to survey for their 'research.' But we got word that three days ago they were spotted off the Kermadec Islands, which means most likely they're going here." He moved his fingers much farther east, almost to the ice-filled indentation of the Ross Ice Shelf. "Which makes sense, because they want to avoid a confrontation with Greenpeace and us. So we're looking at an area between about 160 degrees east to 110 degrees east. From Hobart it's about six days to the Virik Bank. We should be there before Greenpeace."

"So the helicopter is your most important asset in this regard."

"Well, we've also got the stations down here, the French and Australian, on the lookout, plus the supply planes will help out. Senator Lyn Allison is going to bring up a motion in the Australian senate to help us out with intel."

"And if you find them, they just run away; they are so much faster."

"Well, you have to catch them when they're transferring whales to the factory ship. They have to slow down to a speed of five or six knots."

I walked up the hill into the warm bustle of downtown Melbourne.

The quixotic nature of the whole enterprise was becoming painfully clear. I sat at an outdoor café and made a list.

One, the *Farley Mowat* was on her last legs.

Two, the crew were very brave or nuts. Half of them had never been to sea. The pilot of the FIB Thing was not a FIB Thing pilot, but a hang-glider. The chopper pilot was competent but had never flown off a moving deck; nor had he flown a Hughes 300. Watson was taking this old hulk to hunt down and intervene against the Japanese. These are the guys who charged the Nagashino castle with swords because they felt guns were dishonorable, who drank a last toast to the emperor before climbing into a Zero with just enough gas to get to

Halsey's fleet. Their factory ship the *Nisshin Maru* was ten times larger than the *Farley Mowat* and could go 50 percent faster. The *Nisshin Maru* also had five other, even faster ships in escort.

Three, "Aggression on the high seas," as a thirty-year navy veteran had recently informed me, "is grounds for immediate and deadly retaliation."

Four, the odds of finding the fleet were low.

Five, the entire world was against Watson. Even if only through inaction. The Japanese were taking endangered whales out of an internationally established whale sanctuary in antarctic territory claimed by Australia, and Australia did nothing. Nobody else did anything either. Watson was vastly outnumbered.

Six, nobody on the ship gave a damn about odds or allies or anything else but stopping the whalers from killing.

The killing of a whale by the most modern methods is cruel beyond description. An exploding harpoon meant to kill quickly rarely does more than rupture the whale's organs. It thrashes, and gushes blood and begins to drown in its own hemorrhage. It is winched to the side of the kill ship and a probe is jabbed into it and thousands of volts of electricity are run through in an attempt to kill it faster. The whale screams and cries and thrashes. Often, if it is a mother, her calf swims wildly around her, doomed to its own slow death later on. Again, the electricity fails to kill the whale, and it normally takes fifteen to twenty minutes of this torture for the whale to drown and die. Whatever one thinks of whales' high intelligence, the advanced social structures, the obvious emotions and the still mysterious ability to communicate over long distances, this method of slaughter would not be allowed as standard practice in any slaughterhouse in the world.

For the crew of the *Farley Mowat* that was enough.

I thought that Watson was the Anti-Ahab, hunting the ship that hunted the whale. Whereas *Moby-Dick*, the white whale, represented Wild Nature that could not be contained and destroyed—maybe the wild, uncontainable nature in all of us that connects us to other creatures and to the earth—so the mammoth black *Nisshin Maru* was his negative. It was the great, steel juggernaut of civilization, of industry, that was rolling over nature and destroying her. In Melville's day Nature could not be destroyed. In our day she can. The black *Nisshin* and all this vessel symbolizes would succeed in crushing that wildness. Just as the White Whale was in the end indomitable, so now was the factory ship. But Watson, like Ahab, would chase it to the ends of the earth and dash himself against it. His was an archaic mission. Nobody did this kind of thing anymore.

Ahab's old *Pequod* and the *Farley Mowat* were both idiosyncratic boats of long service, at the end of their days. To ship on them as a crew member was a certain act of faith. Each ship had an international crew from every stratum of society that had shipped together for one purpose. Literary critics liked to talk of the *Pequod* as a floating microcosm of democracy, a satellite America living up to its ideals of equality at a time when the country was floundering in slavery and a nascent industrialism that was nearly feudal in the concentration of wealth and the brutality of its exploitation. The *Pequod* was a true meritocracy where the aristocrats of the crew, with the largest shares in the profits, were the three harpooneers—an African, a Native American, and a tattooed South Sea Islander. So the *Farley* could be seen as a microcosm of the best of Globalism—a truly new world order in which daunting pan-national problems such as the deterioration of the biosphere demanded pan-national solutions in which national borders were erased. To tackle climate change, habitat destruction, overpopulation, and mass extinctions, we would surely have to pitch in together as a species and stop thinking foremost about national interest. So with the *Farley* and her single-minded crew of Australians, Canadians, French, New Zealanders, Germans, Swedes, Americans, Brazilians, and Dutch.

I missed my first dinner on the ship. Meals, I was told, were served at 0800, noon and 1800 (six p.m.). I found Geert outside and he offered to show me the union shower facilities. We walked up the wharf chatting like old friends about the day. On the front of his T-shirt was a buxom beauty cupping her breasts and underneath: "Scientific whaling? Right. And these are real."

"Did you do that?"

"Yah, see, that's my trademark. 'Whale Weirdo.'"

We got to the showers, a mobile shed fenced off in an underground parking garage across the street from a high-rise under construction. When Geert took off his shirt I saw that his whole back was tattooed with a sea turtle, and that within its form were other designs, Polynesian and Inuit motifs all having to do with the sea. His upper arms were covered with tribal tattoos, as were his legs. Just like Queequeg in *Moby-Dick*. Over dinner later Geert shared some of his thoughts about the Vipassana Buddhism that he practiced. It was imperative to him to do no harm to anything. That was the first precept. But he was also an expert in kung fu. Buddhist biker vegan black belt children's illustrator eco-pirate.

At the ship, Watson was hanging out with the crew. A semicircle of Shepherds sat on the dock on overturned crates or in folding chairs, and drank beer in the breezy summer warmth. For many of them this was the first real port of call since an arduous crossing of the South Pacific. Watson stood on the main deck of his ship and leaned against the rail and told jokes. He told the one about the man hunting the bear who forces him to have sex. He told about the duck coming into the bar. He told blond jokes, feminist jokes, Irish jokes, jokes about rabbis and priests and Jesus. If it was politically incorrect he told it.

In a rare silence, someone asked if Greenpeace had encountered the Japanese.

"Not yet. Not according to their website. The crew blog today said it was really difficult making banners in the heavy seas with the seventy-degree rolls. Yeah, right." He shook his head. "I've been asking them for months if we can cooperate. With their speed, their ability to locate and keep up with the fleet, and our intervention we could be twice as effective. They won't tell us where their ships are. I've e-mailed John Bowler, the head of their Oceans Campaign; and Shane Rattenbury, who is on the *Esperanza*, leading the expedition—either they don't respond, or they're superior and snotty. 'Sea Shepherd does not meet our threshold of nonviolence.' We've never injured or killed anyone. Not on our crew, not anyone. They can't say that. What I want to know is, how is damaging property used in illegal activities violent?" He tipped his head back and looked out from under his bangs. He was getting warmed up.

"They've never forgiven me for calling them the Avon ladies of the environmental movement. They called me an ecoterrorist. I tried to shrug it off. I was referring to their armies of door-to-door fund-raisers."

"They don't believe in the destruction of property," said Alex, the first mate. He shared a crate with the young blond who had been giving orders all day on the deck. Her name was Kalifi.

"What will they do if they encounter the whalers first?"

"Take pictures. That's all they ever do." Watson drained a can of Foster's and crushed it in his bear paw. "If they were really interested in stopping the whaling they'd work with us. They come down here, take a lot of pictures of whales being slaughtered, and use them to raise a pile of money. They don't stop a single whale getting killed." Watson was almost soft-spoken in the telling, but the heat of his anger radiated from him the way it does from a desert rock at night.

A taxi pulled up. A kid with reddish gelled hair got out. Freckles. A thin, twisted mouth. Tired circles under his eyes. He sauntered up to the ship. You could see he was a scrapper. "Captain," he nodded.

"How'd it go?" asked Gedden. "You have that winning glow."

"Up 300. Justin and Joel are still at the table. I think Justin's up about 800." His name was Jeff Watkins. He had a New York accent. They'd been on the ship a week and every night after dinner he and his two buddies from Syracuse—Justin Pellingra and Joel Capolongo, the "J. crew"—hit the casino in Melbourne and preyed on the tourists at the Texas hold 'em tables. Back home they were committed animal rights activists and professional gamblers who made a good portion of their income playing poker at the Indian casino. They had joined the ship as a squad.

I asked Jeff how he got hooked up with Sea Shepherd.

Jeff stuck his hands in his pockets, took a deep breath and looked us over. "My friend died last month. Twenty-five. Complained of a headache one day—had a tumor about to burst. That was a hard one." Jeff had the most transparent milky skin. His eyes were wet. The way he stood with his hands shoved in his jeans, he looked like a lost kid. "Me and my wife lost our baby, four months' pregnant. I was like, fuck it. I saw that the Shepherds were going out, said I'm going. My two best friends were signed up so I decided to come, too."

Sea Shepherd equals Foreign Legion.

"You ever do anything like this before?" I asked.

"I spent my twenty-first birthday in the hole in lockdown. For destroying a fur coat. A lady just happened to be wearing it when I did it."

I went to bed. I tumbled down the steep steps and pulled the curtain on my snug cabin. I lay down under the musty flannel sleeping bag I had found in a storage closet. I tried to open the porthole, but it was bolted closed. I lay in the close dark and looked up at the glow-in-the-dark poster of the southern constellations taped to the deck head. Somewhere deep in the ship a generator was running.

Somewhere far to the south and west of us the monstrous black *Nisshin Maru* was entering a world of ice, slipping its steel bulk through the drifting icebergs, scattering its catcher and spotter ships to look for prey. They must be very close by now.

From Africa the two Greenpeace ships were sailing east, limited in speed by the slower of the two boats, the *Arctic Sunrise*, an ungainly former sealer with an almost flat bottom that wouldn't go much faster than the *Farley*. It sounded as if they were encountering some big seas and rough going. Soon the *Farley* would be pounding south as well, as fast as she could. No love lost between the three groups.

Nine ships tugged to the bottom of the world by herds of warm-blooded creatures that slipped through the same ice and breathed the same air, that spoke to each other in songs we could not decipher—who must have known, after decades of being hunted, that their season was come.

I think it's safe to say that the whales did not want to be found by their hunters. The Japanese did not want to be found by Greenpeace or Sea Shepherd. Greenpeace evidently did not want to be found by the *Farley*. The *Farley*, which could match no one's speed, would not want to be seen, would want to sneak up on her prey and catch it unawares. Everybody, in some way, was running and hiding. Their agendas, however, were written bold in the media and in their own releases. I knew that Greenpeace was doing a good thing, exposing the sham of "scientific" whaling, and not endangering any lives. Sea Shepherd I wasn't so sure about. Watson was an unknown quantity, and his tactics at times were scary. It seemed a miracle that no one had been killed in any of his actions. I thought as I lay there that Watson would need more than a Little Pecker god and a wooden demon, no matter how ferocious or compassionate, to catch the Japanese. I looked up at the glowing constellations on my poster. The stars were the greenish color of ships on radar. Thousands of them. They swirled and coalesced and none of them looked familiar.

In 1979, after a yearlong hunt, Watson had found the pirate whaler *Sierra* in the waters off Portugal. The whaler was the worst offender in a dirty business. Barred from ports around the world for violating international conventions on whaling and endangered species, and for not paying bills for fuel and provisions, it roved over the globe taking every whale it came across. It had changed its name and flag numerous times in ten years. It was manned by a crew of alleged criminals: numerous countries had issued warrants for their arrest. It often came into port at night, and left in the dark before morning. It sold the meat to Japan. By Watson's estimate, in a decade of poaching whales it had been responsible for the slaughter of 25,000 animals, endangered and unendangered alike. When Watson caught the ship he rammed it at full speed and tore open the hull to the waterline. The next year, after $1 million in uninsured repairs, two unnamed operatives sank it at dockside using limpet mines.

No more whales would be killed by the *Sierra*. Watson got a taste for the immediate, undeniable results of direct "enforcement." He and Sea Shepherd moved on to the *Astrid*, another pirate whaler operating in the Atlantic. They put out a $25,000 dollar reward for its sinking. The ship was sold out of whaling, as the owners felt they could no longer trust the crew. In 1986, Watson commissioned two men to sink two Icelandic whalers in Reykjavík harbor. What did he mean when he told the Australian press he wasn't going to ram the Japanese, but he was going to stop them? That's what he did: in port he sank ships, scuttled them, or blew them open; on the high seas he rammed.

Live by the sword, die by the sword. When I woke up the next morning, disoriented in the close dark, I found myself in my clothes and remembered: ship rules required all crew members to sleep in their clothes, even in port, as it can take only forty-five seconds for a ship to sink, and those extra seconds getting dressed could cost you your life. I thought again about the *Rainbow Warrior*, and the risk of being sabotaged in port, and what a close, dark tomb would be my berth beneath decks.

5

Final Preparations

December 7, 2005

As far as humans know, I don't exist." Heavy Slavic accent, maybe Polish. Those were the first words I heard in the breakfast line that snaked down the long narrow companionway toward the mess. The speaker was slight, stooped, with a gap in his front teeth and lively black eyes.

The Pole was talking to SWAT and to a freckly kid with a turned-up nose.

"Covert action is the way," he said. "I use elite re-breathers, the same the Italian Special Forces use. No sonar signature. I can go for five hours. Actually, we are developing them, my company. Seven kilos. The United States military is twenty years behind."

"I'd like to see a seven-kilo air supply!" announced Freckles in an accent I couldn't place.

"What I'd really like to do," said Pawel Achtel, Mr. Covert, "is go up to the Japanese in the Zodiac and put a magnet with a 'package' look-alike—tell the Japanese they have a cargo underneath and they have half an hour to vacate the ship."

"I've done 4,500 scuba dives in the past five and a half years," declared Freckles. "South Africa, Mozambique, Zambia, Cyprus. I could use that!" South Africa, a touch of Afrikaner—that was the accent. His name was Wessel Jacobsz, and he was our Zodiac tech.

"If you're curious, I'll show you my re-breather," Pawel said. "I have a special dry suit as well that has almost no sonar signature. Of course, once you see it, I will have to kill you."

I got to the counter, which was covered with loaves of fresh bread, jam, oatmeal. Behind the counter were two industrial sinks, and to the left of the sinks a doorway led into the cramped galley. A taut, pretty, forty-something blond in dish gloves scrubbed a stack of bowls and bantered with the crew as they filed past. The cook, Laura, greeted me with a gentle smile. I introduced myself to the dishwasher.

"Allison," she said, and gave me a level, searching look. No nonsense—everything about her, from the short haircut to the visual patdown. She held out her hand, still appraising me, then realized it was gloved and soapy, smiled, shrugged, and said, "I'm the dish girl."

Allison Lance Watson, wife of the captain. In 2003, she and the first mate, Alex, had jumped with knives into a net caged bay in Taiji, Japan and freed dolphins that had been corralled there for slaughter. Every year, the locals slaughtered between 2,000 and 3,000 dolphins, and continue to do so. The slaughter is part of the 23,000 permits issued by the Japanese government annually for small cetaceans. Allison and Alex were promptly arrested. In 2003 Allison had been hauled before a grand jury in the United States in connection with Animal Liberation Front (ALF) and Earth Liberation Front (ELF), two groups high on the target list of the FBI's antiterrorism units. She had refused to testify and been charged, oddly, with perjury. To her, everybody was potentially a fed, or a foreign agent. Watson had written in *Earthforce!*—under "The Laws of Eco-Guerilla Activity"—"Do not participate in any action with any individual unless you have known that person for seven years. Make no exceptions." I do not think he meant an ocean campaign with a ship full of volunteers, but I got the impression that Allison moved in a world where laws like that were inviolable and where wariness meant survival. And she was nothing if not straightforward. When she first met Watson, after a talk he had given in southern California, she shook his hand and said, "My name is Allison Lance and I'm going to marry you."

The mess had four booth tables. A mural of an underwater scene with whales covered one wall. The opposite, starboard side had two portholes. Crew members who couldn't fit at the tables spilled into the green room beside it, another cabin with tables and long cushioned banquettes, and the pride of the *Farley Mowat*—a big-screen, surround-sound entertainment center. One whole wall of the green room was taken up by a shelf of DVDs and videos.

I squeezed in next to the first mate, Alex, who ate quickly, barely listening to the conversations whirling around him. Chris Price, the pilot of the FIB Thing, turned out to be a cross-country trucker from Memphis, Arkansas; he was saying, "Craziest town in the world. In my town they rob the people who buy drugs!"

Alex said, "Yeah, we have a lot to do this morning. We are going to move to another dock and refuel. Hopefully the DC motors are done and we can leave tomorrow. Basically, that means everything has to be tied down."

He had been with Sea Shepherd three and a half years. He admitted that it was a very short time for someone with no previous open-ocean experience to become first officer, but he said that in Sea Shepherd you learn fast. Hair shaved almost to stubble, thin neck and cheeks. The perfect complement to Watson. I sensed the same engine of deep anger humming inside him; but whereas Alex seemed to be consuming himself with his work, the ursine Watson thrived on conflict, fattened on it. Alex said he'd seen Watson lecture in Europe. "I was so impressed with it all I decided I had to join. At the time I had a partner and mortgage payments, and I had a contract—I'm a graphic designer. I planned to join for a year. I came back to Holland and spent two months at home. It drove me nuts, so I came back. Too many campaigns I still want to do."

The buzz of activity on the decks, in the engine room, and in the shops was increasing. Marc Oosterwal, the big welder from Amsterdam, was already crouched on the steel heli deck drilling three-inch holes across which he'd later weld rebar rods to make attachment points for hooks to strap down the chopper. The scream of the bit dropped an octave as he punched through the plate. He reached up for a thick red construction pencil in a pocket on the right shoulder of his coveralls, took a tape measure off his utility belt, and checked the placement of the next hole. A few feet away an assistant in fatigues, who had thick dreadlocks, wire-brushed rust off the edge of the deck so Marc could weld anchors for removable safety railings. The railings would hinge down when the helicopter came in to land to give it an open deck.

On the main deck an Englishman, Graham, who had flown all the way to Melbourne just to help ready the ship, was welding steel plate covers over hydraulic and fuel lines that had been torn off in a stormy Pacific crossing. He explained that the flange of curved steel extending back from the bow and partially sheltering the front of the main deck was called the wavebreaker or whaleback.

Up on the monkey deck—the flat roof of the bridge at the top of the superstructure, which held the three radar masts and the twenty-foot tower of the crow's nest—a compact worker in a cut-off sleeveless flannel shirt and union cap was tying down electrical cables, drilling and screwing down fasteners, and moving to the next with professional speed. This was Sparky Dave DeGraaff, one of the construction workers who had seen the ship from the top of a building and come by. He was a master electrician and shop steward for his union, responsible for the safety and well-being of dozens of electricians on his job. He said he was coming with us.

"I was working on that job up there, the Victoria apartment building. We came over and said, 'What you need?' They said, 'We need an electrician.' I get laid off six weeks at Christmas anyway so it worked out well. I thought, What a great cause. What a great mission. I always wanted to go to see it down there. I thought this is it, I've got to go. I keep two dogs at home, got family taking care of them."

"Not married?" I noticed some gray in the hair straying out from under his cap.

"So far, I'm a free man."

Most of this diverse crew were here because they'd heard Watson speak. Something in the power of the man and in his message compelled people to drop everything—jobs, loves, homes—and follow him to the ends of the earth on missions that offered no guarantee of return. It was the simple moral power of hearing someone say, "I know whaling is wrong. I know killing every living thing in the crippled, gasping ocean is wrong. I don't give a damn about bogus science and all the legal considerations of chickenshit bureaucracies. I know stopping the killing is the right thing to do and I am going to stop it, devil take the consequences." Watson did it because he believed it was the right thing to do. Period. Let the chips fall.

So they came to the *Farley* from every corner of the globe. The trouble was that certain persons and governments did not believe in Watson's initial premise, that what he was doing was the right thing.

Late that morning, the crew moved the ship, slowly, to a more industrial dock less than a mile across the inner harbor. We tied up between a fishing boat—one of its crew stood on his deck and watched our approaching bow with obvious anxiety—and a sailing yacht.

A tanker truck backed in and emptied itself into some of the *Farley*'s seven fuel tanks. The driver left and returned. He made three trips: 110,000 liters of diesel: $56,000, U.S. Watson figured that his 130,000-liter total capacity was enough to go fifty days, if the weather wasn't too bad.

A BMW wagon pulled up and three barrel-chested dudes in surf T-shirts got out and started unloading dozens of wet suits, boxes of Oakley shades. They were big-wave surfers from Tourquay just down the coast who specialized in tow-in surfing—riding waves so big you have to be pulled onto them by a Jet Ski.

The toughest-looking was Maurice Cole, the legendary Australian surfboard shaper. He made boards for many big names, including Tom Curren and Kelly Slater, the seven-time world champion.

"Try 'em on," he said to the black-uniformed crew who filtered out onto the pier to get free gear. "Gonna need shades where you're going."

Maurice and company went up to the bridge to pay their respects to the captain; they asked when we were leaving, and Watson told them about the DC motors we were waiting for.

"These guys took our $7,500 up front, and returned the motors. Didn't touch them. When we asked for the money back they told us to take a flying leap. So now we've got them out with another company."

Maurice, ebullient a moment before, turned suddenly serious. "Who are these guys?"

Watson told him. Maurice registered the name and then he looked unaccountably happy. "Me and the boys will go have a little talk with them," he said. "You'll have your money by tomorrow." He reached down and plucked out one of the five baseball bats stowed in a gap under the drawers of the chart table. He hefted it like an old friend, and then in a fluid, practiced motion he swung it toward an imaginary head and at the last moment chopped it down at a virtual knee. "Never go for the head." He beamed at us. "Kill someone, go to jail. Nah. A love tap at the knee, enough to bring 'em down. Wanna be able to buy 'em a drink in the pub later, eh?"

That evening, one of the best Jet Ski mechanics in South Australia came out to the ship and tuned up the two Sea-Doo Bombardiers pro bono. Gedden squatted on the tarmac of the pier and fidgeted with his new obsession, a donated remote-control helicopter. He planned to use it to fly over the factory ship and video it with a remote camera. Gedden stood back, his corn-silk dreads flying in the breeze like a wind sock, and deftly lifted the little chopper one, two, three feet off the ground. Suddenly it reared like a balky horse and dived for earth with a shattering of plastic.

The next morning the electric motor guy showed up with a wad of money, looking sheepish.

■ ■ ■

We were getting close. The morning of my third day in port, the op-ed page of Melbourne's daily, *The Age*, featured the headline of a long letter: "Australia Must Not Surrender to Japanese Piracy."

Watson blasted: "Has the Australian government decided to prostitute its sovereignty to grovel at the feet of Japanese businessmen or is it cowering in fear of the Japanese economic clout? . . . All it will take to save the whales is for one Australian naval vessel to confront the Japanese fleet in the territory and simply order it to leave. Or are we afraid that the Japanese will simply laugh openly in our face because they are certainly laughing behind our back now."

Bold move, especially with the *Farley* still in port and with one more stop, in Hobart, scheduled on the way south. It was not inconceivable that a pissed-off Australian government could find some pretext to detain the ship. But Australians loved their whales: 1.5 million people went whale watching in Australia every year. Maybe he was gauging public opinion, which was overwhelmingly against the Japanese, and applying just the right amount of heat. It seemed he couldn't resist pricking the authorities wherever he went.

Watson had grown up on the rocky coast of New Brunswick in a village with the lyrical name Saint Andrews by the Sea. It was at the heart of a major lobster fishery. He was one of seven children, the oldest. His father was a cook who worked in local restaurants. He was abusive and beat Watson severely when he was a child. Young Paul was fascinated by the sea, but not by fishing. From an early age he abjured any killing of animals. He used to take his lunch down to the wharf and eat it with the crew of the mail boat.

The town was on a peninsula surrounded by woods. Watson said he spent his childhood running over the forests and brooks, jumping into swimming holes, raising as much hell as a basically sweet kid could raise in the country. He said he and his friends used to play cowboys and Indians with BB guns and real bows and arrows, and once he shot a companion in the butt with a BB because the boy was about to shoot a bird. They used to run out on the trestle and lie on the railroad tracks, squeeze into a low place between the rails and let the trains run over them. They'd crouch just beside the road during a heavy snow and let the fast plows bury them. It was a wonder all of them survived. Young Paul was also "an activist member" of the Kindness Club, a local humane society; he ran the trap lines in the woods and rivers around the town and freed the animals.

When Watson was eleven, his father left the family and went to Toronto. Watson's mother became pregnant two years later, and she went into septic shock at home when the fetus died inside her. She died three days later at the hospital. Watson was there, at home. He watched his mother struggle and his younger sister panic about what to do as his mother slowly died. They finally called an ambulance, but it was too late. Watson never forgave his father for leaving his wife alone with all those kids, for not being there when he was needed, for letting her die.

The family moved to Toronto, to the father's home, and Watson proceeded to spend the next few years running away. "I would get farther every time." Once he was sent to a Catholic home for wayward boys, where the nuns punished his incipient irreverence by making him sit in tubs of ice. It's not hard to see where Watson got his contempt of authority: first from his father, then from the church. All pain came from the ones who made the rules.

The last time he left home, he was sixteen. His father began to beat him and Watson decked him. "It was supremely satisfying," he said. He walked out of the house and never looked back. He went to live with his aunt in Montreal, and at seventeen he hopped a freight to Vancouver, where he met up with a group of environmental activists who were protesting the nuclear tests on Amchitka Island, Alaska— the group that would later become Greenpeace. He took classes at the city college to complete his high school credits and soon began going to sea, first working on a Canadian coast guard weather ship, then joining a Norwegian merchant ship that took him to Asia and the Middle East. He attended Simon Fraser University in Vancouver on and off, never graduating, but continued his education instead on the decks of ships, and in the stacks of books he took with him. He was an ardent lover of poetry, admiring Spenser, Whitman, Coleridge, and he could not read enough about world history and religion. Later he would write, in a passionate repudiation of the crimes human culture had inflicted on the rest of the world's creatures: "To the Earth Warrior, a redwood is more sacred than a religious icon, a species of bird or butterfly is of more value than the crown jewels of a nation, and the survival of a species of cacti is more important than the survival of monuments to human conceit like the pyramids." He would say that all of the arts—music, poetry, architecture—were as dust when compared with the survival of a single species of bird or insect.

Back at the ship they were loading aviation fuel for the chopper. Eighteen fifty-five gallon drums of it. Enough for seventy-five hours in the air. The deck hands wheeled them on a dolly to the port side of the open poop deck and lashed them in with heavy straps. The gasoline was highly combustible. Our little fuel depot was directly under the landing zone of the helicopter . . . in case of an accident, any piece of flying debris . . . I didn't want to think about it. On a voyage like this, you just couldn't sweat the small stuff.

After dinner, on the night of December 9, the cry went through the ship that we were leaving the next morning and that everything had to be "tied down." All personal items in the cabins had to be stowed and secured. All supplies, all tools, all utensils—everything, down to the last spoon and coffee cup—had to find its place, and its place had to hold the item securely against a thirty to forty degree roll, perhaps more. Squads went through the ship with rope and carabiners, straps. They went into the fish hold and lashed the FIB Thing into submission on its frame. In the same hold they cinched and bound the long heavy beams and rods of steel—twelve-foot lengths of four-inch T-bar and angle iron, rebar—and the stacks of lumber used in framing and repairs. On either side of the poop deck aft, under the heli deck, they strapped down the rows of steel drums, jet fuel to port, lube oil to starboard. Books and videos snugged into their cabinets. Kitchen knives, into their drawered slots. Navigation rulers onto their racks. Sacks of turnips and potatoes were lowered into dry storage, a hole aft of the mess and down a vertical ladder to the lowest deck.

Crates and crates of boxed soy and rice milk squeezed into compartments under every bench seat in the mess and green room. Out on the low open main deck the newly frisky Jet Skis were lashed into their frames. The surf mechanic had at first opened the engine compartment, taken one glance and stretched his knotty arms to the sky in a kind of supplication. Farther forward, the Zodiacs were tarped and lashed neatly into their own cradles. The efficiency of the old hands, the ones who had been on board for the long crossing from the Galápagos, was impressive. Gedden, Colin, Inde, Ryan; Kalifi and Steve: they tore from one end of the ship to the other and in two hours it was done. You could have picked up the ship and shaken it upside down like Pierre's lion, and the only thing falling out would have been a few Australian coins and a rubber duck. Maybe a small poster of the southern constellations.

I went down to my cabin forward and put boots into the locker and laptop into a drawer of the desk that was fitted with a shock cord running from one leg to a wooden cleat nailed to the top. Then I met Ron Colby, one of the documentary filmmakers, and we strolled down the pier, heading for the bars and restaurants, for a last hamburger before the vegan wasteland of Antarctica.

Ron is a veteran Hollywood producer and director, in his sixties but seeming much younger, lean, tough, sardonic, with a resonant voice and a way of observing and listening that disarmed everyone he met. He knew everyone and had done everything, and wouldn't talk about it unless you tortured him into it. He produced *The Outsiders*, among other films. We walked into the first Greek restaurant at the head of the wharf, and were seated next to three hardened-looking young men in black pirate T-shirts—Ryan, Steve the ex-soldier, and Colin of the SWAT suit. We slid the two tables together.

"We will eat all the meat they put in front of us," Colin smiled. Turned out he was a butcher from the Inter-Lakes district outside Winnipeg.

Steve said, "On the way from the Galápagos we pulled out an old tin of roast beef to cook and Inde and Laura cried."

We ordered souvlaki, burgers, steaks.

That night I took welder Marc's four-hour security watch at the gangplank. He was really taken with a local woman; it was their last chance to spend time together.

"You got lucky," I said.

"Yeah, I don't get lucky like this. Good to have a bit of luck." His broad face with the lamb-chop sideburns cracked open in a big smile. He picked me up by the shirt like a bag of laundry and set me down. "T'anks," he said and jogged down the gangplank to find his girl.

I sat at the top of the gangplank in a chilly onshore wind and watched the harbor's seagulls enact a bizarre feeding adaptation. They flitted in and out of the arc lights along the dock, catching insects. They weren't very good at it. They tried to snap and veer like swallows and overshot and missed. I could see the panicked bugs. In swallow country the insects would be already dead, but here their odds looked about fifty-fifty. The gulls swooped and pulled up and stalled and fluttered awkwardly. They hadn't evolved for this kind of thing. But there must have been enough food, because they kept at it for hours. They seemed a bit like Sea Shepherd, doing the best they could with what they had and not too worried about how it looked.

A mile away, at the mouth of the inner harbor, was a high bridge over the River Yarra. Atop the stone tower, very tall, were floodlights that shot a straight beam 500 more feet into the night sky. The swath of light was filled with moving birds. Hundreds of seagulls climbing one narrow beam, gyring, diving, climbing again, feeding on the insects also drawn to the light. An almost solid column of living flight. I tried to keep track of the highest bird, a white cipher against the dark, much too high, I thought, to be catching bugs—up there, it seemed, for some other reason.

The next morning I was woken by the deep throb of the *Farley*'s 1,400-horsepower diesel engine, and within two hours we were casting off ropes and turning south for the bridge and the bay.

6

Salt

The *Farley*, once loosed from the mouth of the River Yarra and out into the stiff southwest wind of the open bay, began to roll with a gentle, easy rhythm you knew she was born to. After two years of preparation, after a previous Antarctic campaign that was a bust, after patchwork repairs all over the ship, after reading the Greenpeace blogs from the *Esperanza*, which sailed closer and closer to the whaling fleet while the *Farley* was held in port by one maintenance issue after another, Sea Shepherd had at last loosed her on the tide. She was gaited for the gales and heavy weather of the North Sea; and as old as she was, the *Farley* bore south toward the rip in the South Channel with vigor. She was heading almost due south, toward the Southern Ocean and the worst seas on earth.

We got to the narrow rip between the Victoria headlands just after 1300 (one p.m.) on a nine-knot outgoing tide. We could see the surf pounding on the outer beaches. A dangerous spot strewn with wrecks. Improbably, just as we were aiming for the slot, a Jet Ski bounded up alongside and the two hooded riders gave us a big thumbs-up and wave—our sendoff by the surfers of Victoria. Then the *Farley* banged through the steep chop of the channel, taking water over the bow and a gush through the main hatch that flooded the long companionway. She nosed out of the slip and into the relative calm of the Bass Strait and the Tasman Sea. This place of peace, the pilot told us, was called Sara's Bosom.

"Who was Sara?" asked the young second officer, Hammarstedt.

"She had an awfully big bosom by the looks of it." The pilot winked and wished us good luck as his orange pilot boat pounded alongside and he went over the rail. They peeled off for the bay. Watson set a course of 131 degrees, southeast, to Tasmania. That put the southwest swell directly on our starboard beam and the eight-foot seas tipped and rocked us like a swinging cradle. Geert and I lashed a couple of beat-up wooden chairs next to the life raft pod on the stern and watched the wake through the stern chains, and the sun lower behind us. It gilded the water as it expired. A twenty-knot wind blew the tops off the whitecaps and rained them downwind, churning blooms of azure out of the dark water. The sky was bell clear. A half-moon rose out of the sea.

Watson tore into Greenpeace on the bridge.

"Greenpeace has a policy called 'bearing witness'—some sort of weird—" Watson looked out the bridge windows at the bow of his ship tipping and lifting on the next swell. "It's a Quaker thing. I guess that's all they had to do in World War II is bear witness to the concentration camps. Saw a lot of Jews bearing witness, didn't do much good. On the back of the union shirts here it says, 'If you don't fight, you lose.'"

He then read from Andrew's blog, on board the *Esperanza*, December 5.

"The last twenty-four hours on the *Arctic Sunrise* have been a bit of a nightmare. . . . It wasn't until 0400 that we fell into a hole in the ocean and achieved our first seventy-degree roll—" Watson looked around at his watch officers. "Jeez. Seventy degrees. Well, we've never done more than a thirty-five- to forty-degree roll on this ship." He stood out of his captain's chair and pointed at the little roll gauge tacked up over the center window. It was a calibrated arc filled with fluid like a level. "Here, this is sixty. That's virtually impossible." He continued with Andrew. "'Last night was also pretty bad. I found myself mopping up soy sauce in the galley at two-thirty in the morning after a big glass bottle of it took a dive. Work went on today though, as it always does. The crew painted some fine banners despite the adverse conditions. Magnets were used to hold the materials down. Steel floors (A. K. A. "decks") are handy that way. There was nothing holding the banner painters down, however, resulting in a certain amount of sliding around.'"

There was something in Andrew's tone. Sounded pretty rough down there.

At dinner, in the mess, a free seat was uncharacteristically easy to find. Hardly anybody was eating. Most of the crew who were there picked gamely at their tofu goulash and stared vacantly at the tabletop or the wall. Seasickness can be completely debilitating. The crew members who were suffering the worst looked as if they'd just as soon get shot. On the aft deck two-thirds of the J. Crew, Justin and Jeff, sprawled in a row of wrought-iron theater chairs bolted to the deck just aft of the main stack and facing astern. Others stumbled out of the hatches and went straight for the rails, and if they remembered to go to the downwind side they were lucky.

Side rolls were the worst. Most of the bunks were oriented forward and aft, so even the crew who took to their beds had to brace themselves in and hold on to keep from getting thrown to the deck.

Our registered nurse was philosophical. "It'll pass, I'm sure," she said. "I hope." She didn't look so good herself. Kristy Whitefield was an emergency room nurse in Melbourne. She was five feet in her bright red socks, maybe ninety pounds. Her wild dreadlocks added another couple of inches. She wore rainbow-striped leg and arm warmers she had made herself. She moved around the ship like a brilliant jungle bird whose nest is on its head. "I didn't bring motion sickness medicine," she said. "You just can't bring enough." Another animal rights activist, she told me she did "hunt sabs"—actions sabotaging hunters—during the opening days of Australia's duck season. Before dawn she and her compadres waded into the frigid waters under the loaded guns of the hunters and banged pots and pans and yelled. She'd been punched, held underwater, threatened at gunpoint.

On the way forward, I passed a lanky figure leaning against the superstructure smoking a cigarette with the relaxed pose of a man leaning on a lamppost. There was an easy grace in the way he shifted his weight with the roll so that he didn't seem to be moving at all. He wore dark blue coveralls and red ear protectors propped on his closely cropped head. He watched the waves coming out of the dark onto the starboard beam and heave under the ship. The bow made a rhythmic sifting wash against them.

"Nice night," he said.

"For some of us."

He laughed softly. "They'll straighten out." The way he said "out" I knew he was Canadian. He crushed the butt against the steel and put it in his pocket. "Best weather we may get all trip."

He introduced himself, Trevor Van Der Gulik, the chief engineer. He was thirty and had been coming on campaigns with Sea Shepherd since he was fourteen; he was Watson's nephew—his mother was Watson's sister. Trevor lived now with his Iranian wife on the same rocky coast of New Brunswick where Watson had spent much of his childhood. He had worked on many ships, including a motor-sailing brigantine, and most recently had been a supervising engineer in the largest private shipyard in the world, in Dubai. Often he'd had 500 mechanics under him. They worked on diesel engines three stories high.

"Good money," he said, and lit another Marlboro. "I should have stayed, but then I wouldn't get to do this." He smoked, and we watched the stars swing toward and away from the black horizon. "I live for this," he said.

Trevor was one of the three paid officers on the otherwise all-volunteer ship. The others were the second engineer, his assistant: and the first mate, Alex. They didn't get much, just a stipend to keep them from going in the hole on a campaign. Trevor said he went to school for his marine engineering technology degree when he was nineteen. He'd already been at sea on and off for five years, so it was easy—all he had to do was buckle down and learn the math and thermodynamics.

"It's all about the transfer of heat and energy. Energy cannot be created, which makes sense to me. It's always moving from available energy to unavailable energy. But that doesn't mean it's unavailable to the universe. Entropy is the measurement of unavailable energy. I love entropy."

Trevor rocked off the bulkhead and stood square on his feet. "Working in the conservation movement so long, it's all about how we use energy in society, how much we use per person per day. Back in preindustrial times, we grew our food, had our own farm—or were hunter-gatherers. We may have used an ounce of coal or two in one day. Now we'll use five liters of gasoline to hop into the car and go get cigarettes. May use five pounds of coal per day. Only so much energy on the planet. More we use, the less there is. Inevitably we will have earth death, because there is no more available energy on the planet. That is inevitable. Up to us to try to make sure it's as slow as possible. Entropy." He ground out the cigarette butt. "I've got to go take care of some right now. In my Entropy Room. So we can burn a bunch of available energy to do a good thing." He went through the hatch that led straight down to the engine room.

At some point in the night we would cross into the "roaring forties." A year earlier the Bass Strait had turned back fifty-seven out of 116 yachts that attempted the Sydney-to-Hobart race. A dangerous stretch of water. In 1998 the race claimed six lives. Well, they were sailboats.

December 11, Sunday, 0854
40°12′S, 147°25′E

The mountains of Flinders Island visible at daybreak off port. Australian gannets following the ship, swooping over the wake. Boobies and seagulls. Seas calm as we enter the roaring forties, still rolling on the beam. Few people at breakfast, the collective nausea having tightened its hold. By the doorway to the mess was a dry-eraser board and cork board. Tacked to the cork board was the following announcement:

If you are unhappy about the way this ship is run, it is simple . . . get off! Discord usually starts with a couple of people and snowballs. This will be our most aggressive and ambitious campaign ever. Therefore we all need to be in sync. Thanks—Allison.

The Enforcer. Seemed a bit late, the part about getting off, unless people walked the plank, or marooned themselves in Tasmania. That was not out of the question. Watson had abandoned crew before, in foreign ports, without hesitation, when he thought they lacked the proper discipline or commitment and jeopardized the mission. The import was clear. Dissension would not be tolerated. Watson liked to quote Captain Kirk of the Starship *Enterprise*: "When this ship becomes a democracy, you'll be the first to know." The first whiff of grumbling, like a thread of smoke from the forest duff, had been stomped.

The forty-four-member crew was divided, roughly, into four areas: bridge, deck, engine, and galley. The bridge and engine room were the only divisions that had scheduled rotating watches, because someone had to be at the helm and manning the machinery 24/7. For the engine room, this meant that Trevor was down there almost every hour when he wasn't sleeping or eating.

The bridge was served by three four-hour watches or teams that rotated around the clock. Captain Watson took the watch from eight a.m. to noon; the second officer, Hammarstedt, took noon to four p.m.; and Alex, the first officer, took four to eight p.m. Then Watson again, eight p.m. to midnight, and the rotation began again. Day in, day out. All ships use twenty-four-hour or military time, so that meant 0800 to 1200, 1200 to 1600, 1600 to 2000, 2000 to 2400, 2400 to 0400, 0400 to 0800. Each of the officers had two quartermasters on his watch to steer the ship when steering was required—as through ice fields. Often the steering was done by autopilot, while the quartermasters monitored the radar, the radio, and the ship's position; plotted the position every hour on the chart; and entered it hourly into the ship's log, along with speed, current heading, and relevant notes about weather, fuel, actions, etc. And kept a sharp lookout for ice, whales, and ships.

Especially ice. The waters off the coast of Antarctica are exceptionally treacherous, for not only are the winds pouring off the continent the strongest on earth, and the seas the most volatile—three oceans meet, and there is nothing around the whole girth of the continent for thousands of miles to impede their fury—but in the Antarctic summer the calving glaciers and deteriorating bergs litter the waters with drifting ice of every description. Most perilous for the *Farley* were the "growlers," large chunks of ice that floated just at the surface and were hard to see until you were right on them, especially in fog, especially in heavy swells. The bridge watch, once we got down to the ice, would have its hands full. A pair of good marine binoculars sat in a small cabinet on the port side of the bridge, and would be used constantly by the quartermasters.

Spotting ships on radar in icy waters was another demanding job. It took special vigilance and skill. The main radar, set into the bridge console just to the port side of center, could be set out to sixty-four miles, but it could discern discrete objects accurately only out to thirty-two miles, and with any usable clarity only out to about twenty miles. Also, it is very difficult to tell an iceberg from a ship.

The engine room was the concern of Trevor, the chief engineer: his assistant, the second engineer, Willie Houtman: and four engineers. This allowed for continuous rotating shifts of two men, as needed. The engine room had two decks, or levels, and was a universe unto itself. The 1,400-horsepower MAN-made diesel they tended had seven cylinders and was twenty-one feet long. The men who worked on it lived in a cloistered din of exploding pistons, racketing valves, thrumming generators—pumps, hydraulics, cams, gears, all clamoring at once, all the time. All these engines and motors and pumps dripped oil, water, fuel constantly, so one of the engineers was pretty much a dedicated swabber. Trevor could stand nothing less than a spotless entropy room: he hated nothing more than a greasy handrail or metal step.

It was monastic down there, and as with many such insulated fraternities, it fostered in the engineers an esprit de corps and devotion unlike any other watch on the ship. On a quiet evening you'd see two or three of them on deck just outside their hatch, ear protectors propped on heads, grease-smeared and pale under the grime, smoking cigarettes. The smoking porch was on the starboard side, sheltered by the superstructure and the foot of the outer stairs leading up to the bridge wing. A life raft pod, about four feet long, was cradled there against the bulkhead, an ideal bench seat with a view of the ocean over the starboard rail. You got a sense that for Willie, a twenty-two-year-old Kiwi, who joined the ship when he was eighteen and had made the *Farley*'s engine his university, the engine room was a sanctuary, that he'd rather be there than anywhere else. He wore nothing but stained and tattered rags, and his dreadlocks were a mane of raggedy hair, so he moved in the engine room in a kind of camouflage. I began to think of him as Willie Tatters.

Trevor could stand in the middle of the thunderous din, eyes closed, and pick out the smallest tic, a hitch in cadence, or a shift in key, and know where to go and what to tweak. On the bridge, in a storm, in the middle of a raucous conversation, he could go suddenly still. "Lube oil separator," he'd say and disappear. In the same way, he could be down in the engine room and have a good idea about decisions being made on the bridge: he'd be working figures with a pencil and a book of fuel tank tables and look up and say, "They just changed course, about twenty degrees off the weather."

Any aspiring engineer who showed up with an iPod and headphones was shown the hatch. "Down here, you learn with your ears," he said.

All things deckwise were under the command of the bosun. Kalifi Moon Ferretti-Gallon was the young woman I'd seen yelling orders and running the davit crane in port. She had plucked eyebrows and a fine aquiline nose, and was fond of wearing pajama bottoms and pink thong sandals adorned with sequins. Her curly blond hair was uncontrollable. She was twenty-two, was partway through college, came from Montreal, where she had most recently been working as a barmaid and was not sure what to do with her life. Her father, Gary Gallon, started Greenpeace with Watson in 1972, and was a large presence in the Canadian and international environmental movement. He had died two years earlier, at fifty-eight, of cancer. Kalifi said, "Only two daughters. He grew up teaching me about engines. If he knew I was bosun, in charge of the deck, saving whales, he'd be really proud." She had been on the ship for six months, since Bermuda—this was one of the longest continuous stints among the present crew—and she had worked her way up swiftly into the cadre of officers. Her current boyfriend was Alex Cornelissen, the first officer. This caused some consternation among the crew, and suspicions of favoritism, but Alex swore he gave his girl a harder time about ship's business than he would have given any other bosun.

By a happy coincidence, Kalifi's best friend, Emily Hunter, twenty-one, of Toronto was also on the ship. Emily's father was also one of the pioneers who started Greenpeace with Watson, and also a major figure in the international environmental movement. He also died of cancer, just six months before the *Farley*'s departure for Antarctica, and Emily had his ashes with her to be spread amid the ice and the whales of the Antarctic seas. She said, "I know the trip is dangerous. I don't want to die, of course; I'm very young. But if I die helping to save a whale, that would be OK. That would be a good way to go."

Under Kalifi, in command of the deck, was the bosun's mate, Colin Miller, the very focused former butcher from outside Winnipeg who never went to work on the ship without his SWAT vest. The knife he wore on the vest was an Ocean Master Beta Titanium, which can take 250,000 pounds per square inch of pressure before the blade snaps. He also carried headlamp, tools, knife, radio. At home, on Sunday nights, Colin does serious Close Quarter Breach tactical training—for fun. "Sweep and clear kind of thing, two teams." On the first day at sea, Colin took me to his cabin just aft of mine and opened up his laptop to show me some of the weapons he trains with. The screen filled with a very efficient-looking Steyr AUG 5.56, black, fully automatic assault rifle paint gun. "Red dot site. Touch here, the barrel, and the tac site will come up. But I use a G36, like a submachine gun."

He also trained in Malaysian knife fighting. The style is called Kali ("Did you ever see the movie *The Hunted*?") He uses a curved knife called a Karambit with a finger ring for extending reach. "There is no good knife fighting. There's nothing beautiful about knife fighting," he said. The style, which uses a lot of hand blocks, originated with slaves fighting in the Philippines. "Being a butcher—" he said, "the transition from cutting up steaks to cutting up a person is not that different."

Colin was universally respected. He never asked a hand to do something he wouldn't do; he was even-tempered and patient in teaching the greener crew. He never got irritated and he never complained. He was a perfect team player.

There were eight deck hands. Three of them were the J. Crew, the professional gamblers from Syracuse who were now on the aft deck retching. The deck hands were responsible for rope handling; docking and casting off; launching and manning the attack Zodiacs; securing the Jet Skis and inflatables; tying down everything on the decks and in the holds; handling the anchors and chains, davits, and knuckle boom; and painting, fiberglassing, and maintaining almost everything, including the general cleanliness of the *Farley*. They made up the fire suppression teams. They were the backbone of the crew. Their headquarters was the bosun's locker, an all-purpose workshop with two workbenches just off the main deck at the front of the superstructure. The aft bulkhead or wall of the bosun's locker was solid with shelves of milk crates full of cord, rope, cable, shackles, nuts and bolts, chain. A couple of fifty-five-gallon drums were lashed into the corner and filled with scrap wood, steel, rubber.

Laura Bridget Dakin, twenty-two, holder of three passports— United Kingdom, Bermuda and Australia—horse trainer and champion endurance rider, who embarked for Antarctica with nothing but flip-flop sandals, a flowered skirt, and a lip ring, was the chief cook.

She had a second and a third cook working beneath her, and a galley assistant, who happened to be Allison. Cooking three vegan meals a day for forty-four people, including fresh breads and desserts, on surfaces that are often rolling twenty degrees or more, might seem a daunting task, especially for someone who had never done any kind of institutional cooking before embarking on the ship last June in Bermuda. But Laura was not easily ruffled. In Victoria, Australia, where she spent much of her childhood, she regularly rode brutal 100-mile endurance races on horseback, many miles of which covered rough and dangerous terrain in the predawn dark, and she often won her weight class. The reports that the Japanese were beginning to kill whales fired her blue eyes with a ferocity that was unnerving.

Her second cook, Roberta Kleber, was an enigma. She came from Brazil, but her accent was German, as were her name and appearance: high snowy forehead, small straight nose, hollow serious blue eyes. Her town in southern Brazil had 6,000 citizens, all of whom spoke German. Four generations. She was very quiet and a hard worker. She was hitched up with Gunter Schwabenland, quartermaster and Sea Shepherd photographer, who came from another one of these segregated towns of 30,000 German-speakers, also in the south of Brazil.

The third cook was Casson Trenor. He was the only person on the ship who had an iceberg's chance in hell of beating Watson at Trivial Pursuit, or solving one of the captain's arcane riddles. Before Watson got addicted to Texas Hold 'em, his favorite after-dinner activity was to pull out the battered box from the game shelf in the rear of the mess; pluck out a brick of stained Trivia cards; pass them to Hammarstedt, who would read the categories and questions; and then whip the pants off of all comers in Speed Pursuit. Watson had an astounding head for historical, scientific, and geographical minutiae. Which was evident when he lectured the beleaguered young quartermasters of his watch on the taking of Persia by the Mongols, the taxonomic relationships of prehuman primates, or the deficiencies of the medieval pope Boniface VIII. In his book *Earthforce!* Watson had recommended just making up the facts in a pinch; but you couldn't make up the headwaters of the Volga or the name of Queen Elizabeth's lover—not in the split second it took Watson to shout the name. The only one who sometimes beat Watson was Casson. He was lean, slouching, handsome, twenty-six, recently of Monterey, California. He held a master's degree in international environmental policy, and a chef's certificate from the esteemed Pacific Culinary Institute in Vancouver. He spoke fluent French, Arabic, Spanish and passable Japanese. He also knew Latin. He had worked for the Conservation Strategy Fund advising developing countries in the South Pacific how best to spend their limited conservation funds. For those island nations, because they are so bounded and their natural resources so circumscribed, environmental conservation is more obviously a matter of economic life and death than for almost anyplace else in the world. Introduce one giant African snail that kills off 70 percent of your cassava, and your economy is wiped out—and your livelihood, your health, your very lives are in danger. Casson had also worked for Seafood Watch, which puts out the wallet cards that list which fish species are environmentally sound to eat. It's a very short list.

Casson believed, along with many of his marine biologist colleagues, that the oceans are on the verge of total ecosystem collapse.

Consider: 90 percent of the large predatory fish, the tuna, swordfish, marlin, sharks—vital to the health of ocean ecosystems as we know them—have vanished since 1950. The World Wildlife Fund recently announced that the world's fish stocks are on the verge of extinction. All the world's fish. Right now, half of the world's coral reefs are dead or dying.

"If the oceans die, we die, too—" Casson, at a table in the mess, chopping turnips, saying that much of the destruction is caused by bottom trawlers, the industrial dragnets that scour and scrape wide swaths of sea floor, taking everything—the octopuses, the sea turtles, the mammals, the crabs, the urchins—and damaging the reefs. The weighted nets are dragged in and the target species are thrown into the hold, while all the rest are chucked overboard, dead or dying. Wasted. Casson telling us that bottom trawlers scour an area twice the size of the United States every year. That between the longlines and bottom trawlers 7 million tons of thriving ocean life is tossed as bycatch annually. Hundreds of thousands of sea mammals—seals, dolphins, sea lions—are slaughtered and dumped; 100,000 albatross. So many of the widest-winged pelagic birds that they are now also threatened with extinction.

"Bottom trawling is like bulldozing the forest, killing everything in it, to get at the wild turkeys. That's what it leaves, a smoking waste. Longlining isn't much better. It's indiscriminate. Tens of thousands of sea turtles are killed and thrown overboard as bycatch each year. Seabirds, especially albatross, dive on the baited hooks and get dragged under and drowned. It's insanity. These are wild animals. It'd be like going onto the African savanna and shooting every living thing, mammal, bird, that you see."

Laura chimed in: "Have you seen a dolphin caught in a trawl net? Did you know they have been shown to call each other by name?" She got up and retrieved a sheet of paper. "Here, Peter," she said in her genteel British accent. It was a printout of a forwarded e-mail:

If you read the front-page story of the *SF Chronicle*, you would have read about a female humpback whale who had become entangled in a spiderweb of crab traps and lines. She was weighted down by hundreds of pounds of traps that caused her to struggle to stay afloat. She also had hundreds of yards of line rope wrapped around her body, her tail, her torso, a line tugging in her mouth. A fisherman spotted her just east of the Farralon Islands (outside the Golden Gate) and radioed an environmental group for help. Within a few hours, the rescue team arrived and determined that she was so bad off, the only way to save her was to dive in and untangle her . . . a very dangerous proposition. One slap of the tail could kill a rescuer. They worked for hours with curved knives and eventually freed her. When she was free, the divers say she swam in what seemed like joyous circles. She then came back to each and every diver, one at a time, and nudged them, pushed them gently around—she thanked them. Some said it was the most incredibly beautiful experience of their lives. The guy who cut the rope out of her mouth says her eye was following him the whole time, and he will never be the same.

"How can they kill these creatures?" she asked. "The earth is 75 percent ocean. Why is this the only ship out on the ocean enforcing environmental laws? One ship protecting the ocean?"

Recently I had been in correspondence with Dr. Sylvia Earle, the legendary oceanographer and marine biologist who is explorer-in-residence at the National Geographic Society. Casson had echoed, almost issue by issue, the great scientist's alarm over the crisis of the oceans. Earle wrote:

"In the latter half of the 20th century, ocean exploration yielded stunning new discoveries. In a few remarkable decades we . . . began to understand . . . how the ocean governs climate and weather, shapes planetary chemistry, generates most of the oxygen in the atmosphere, and absorbs large quantities of carbon dioxide. We gained new appreciation for the ocean as the cornerstone of our life support system, not just because of the abundance of water, but because of the living systems in the water that over eons have changed the nature of the planet from a hostile place to one that is hospitable for us and for life as we know it.

In half a century, we learned more about the ocean than during all preceding history, but at the same time, we lost more. In fifty years, half of the world's coral reefs have disappeared or are in a state of serious decline, ninety percent of the large predatory fish have been taken—tunas, swordfish, sharks, grouper, snapper, halibut, cod and many others. Destructive fishing practices—trawling, longlining, driftnetting—have depleted not only targeted species, but also devastated diverse ocean ecosystems and killed millions of sea birds, marine mammals, turtles and countless other creatures as throwaway "bycatch." In coastal regions around the world, more than 150 "dead zones" have developed as a consequence of noxious substances we have allowed to flow into the sea. Excess carbon dioxide released into the atmosphere and absorbed by the ocean is increasing acidity of the water that in turn poses threats to creatures dependent on calcium carbonate for their survival. This includes corals and many other animals as well as a large number of the small photosynthetic organisms responsible for generating food and oxygen that drive planetary chemistry.

The good news is that half of the coral reefs are still in good shape and ten percent of the big fish remains. Sea turtles are not all gone, and some albatrosses still grace the skies above the world's oceans. Best of all, people are becoming aware of the significance of the ocean to their health, their prosperity, their security, and most importantly, their survival in a universe where there are no immediate alternatives to sustaining ourselves on this water-blessed planet. With knowing comes caring, and with caring, there is hope that we'll find an enduring place for ourselves within the natural systems that sustain us."

Once asked what she considered the greatest threat to the world's oceans, Earle responded: "Ignorance, lack of understanding, a failure to relate our destiny to that of the sea."

There was cross-pollination between animal rights activists and conservationists on the *Farley*. Between Casson and Laura. They were all here together on this single battering ram of ship, which was hurtling like a very slow arrow toward its target. The marriage of the two groups, who are often seen at cross-purposes politically, was epitomized in the marriage of Watson to Allison Lance, who, together, ran the ship. "He is a conservationist," said Allison. "He thinks 700 years in the future. I am concerned with each and every living thing that cannot speak for itself."

She was a strict vegan. "He is barely a vegetarian," said Ron, the producer, who had known Watson for twenty years, since long before Allison came into the picture seven years ago. Watson had adapted the diet and now saw the link between animal rights and conservation. "I think the consumption of domesticated animals is a major contributor to environmental degradation," he said. "And seafood—where there is simply not enough fish in the ocean, demand has exceeded supply. More than 50 percent is fed to terrestrial animals. We've turned chicken, pigs, sheep, and cats into major aquatic predators."

In 2003 he got himself elected to the board of directors of the Sierra Club with a push of support from animal rights and antihunting groups. He terrified old guard of the august organization with the prospect of a coup by animal rights supporters.

There are many ardent conservationists who believe that all whaling should stop, but who also think animal rights people are kooks. Watson said, "I don't give a damn. Opposing hunting, opposing factory farming is the right thing to do." I wonder, though, on the *Farley*, how the groups are apportioned, or if there is any real division, aside from a difference in emphasis. Judging by the surreptitious meat eaters at the Greek restaurant in Melbourne the night before departure, I think there is. How many on the ship are vegan? How many are vegetarian? How many eat almost anything?

And does it matter? They were all on board, heading south to stop the whalers. All risking their lives in this one goal. Did it matter if one chose to eat an egg? Or honey? Or elk tenderloin?

Did it matter that on board Greenpeace's marine conservation ship, the *Esperanza*, the crew ate fish?

Bridge, engine room, deck, galley. The fifth division on the ship was the fourth estate. The media had a strong presence on the *Farley*. Of forty-four people on board, seven were journalists of one form or another. There was Paul Taggart, the slight, 120-pound, twenty-five-year-old war photographer from Tulsa, Oklahoma—now of Brooklyn, New York. He wore wire-rimmed glasses and was clean-cut, skeptical, kind of quiet—but with nerves of Damask steel. Blood like ice water. The Peter Parker of the ship. The kid spent 300 days a year on the road in places like Baghdad, Monrovia, and Banda Aceh, shooting the most horrific scenes. He works for the World Picture News photo agency in New York City. International hot spot—he grabbed his cameras and went. He was kidnapped in Iraq by Shiite militants from Sadr City, kept his ever-loving cool, and was released after three days. On the ship, no matter how dire the engagement or how bad the storm, he was unruffled. He climbed into the chopper with his cameras. Into the lead attack Zodiac. He got the shot. He didn't even need to change into a Lycra suit.

"It's kind of nice," he said, "shooting an assignment that's not full of dead people." He had a quiet, deriding laugh and a way of darting his gimlet blue eyes sideways when he heard something over the top. Which was all the time on the *Farley*. He and the pilot, Aultman—partly because they spent so much time in the chopper together, and partly because of their congenital skepticism—had an affinity and a bond.

There were three filmmakers on the boat, shooting documentaries. Mathieu Mauvernay, thirty, of Paris, was one of the first crew members I encountered on the *Farley*, barefoot in a bucket, stomping grapes. He was the brother of Nicolas Mauvernay, the famous producer-director of *Winged Migration*. Mathieu used to be an environmental lawyer—for the bad guys.

"I think Paul saved my life. I been defending the big oil companies. After the *Erika* oil spill in France, five or six years ago. Total Petrofina, we made them won. I was a bit sad about that. The beaches in the west were devastated, but nobody from the company went to clean them. It made me sad. I went to work for Paul. I became an eco warrior with the Sea Shepherd." He grinned. He had a Gallic beak of a nose. He owned a set of yellow foul-weather bib overalls that he wore twenty-four hours a day. In March 2005, he had accompanied Watson onto the ice in Newfoundland in an attempt to stop the Canadian seal hunt, and he shot a powerful documentary that is very difficult to watch. Each year, since 1994, weeks after the harp seals give birth, Canadians stalk the ice with hackepicks, long poles with sharp picks on the end, and beat up to 350,000 six-week-old seals to death. Canadian fisheries claim that the seals are responsible for the decline in the cod fishery. Peter Hammarstedt, the beanpole second officer, said, "The Canadians can't blame themselves for the decline in fish stocks, so they blame the seals. They claim the harp seal is killing all the cod. Just 200 years ago there were 41 million seals, now there are only about 2 million to 4 million. They certainly weren't killing all the cod!"

The second videographer, producer Ron, had been shooting a documentary of Watson on and off for years. On the ship, in the green room, there were thousands of DVDs and videos, most of which were stamped with "For Your Consideration"—for one Academy Award or another. Every time Ron came to the ship he dropped off sacks of them. Ron was tough. He was one of the few people on the ship who didn't get sick. Nothing pissed him off but a man killing a helpless sea mammal. On most days at sea he parked himself in the one armchair bolted to the port side of the bridge. His video camera was on the deck beside him. He read *Walden*.

Kristian Olsen was the third videographer. He was thirty-five but appeared much younger, with boyish good looks and shaggy straight blond hair. He was built like the hockey player that he was. He lived in Vancouver, where he worked as a video editor. He was shooting Sea Shepherd for a Canadian film company that was putting together another documentary on Watson.

Pawel Achtel, Mr. Covert, with the special ops, no sonar signature dive outfit, was the other crew member who often had a camera on his shoulder. He landed a deal with ABC Australia to cover the campaign. In exchange for exclusive use of Pawel's first images it supplied the ship with a new satellite transceiver, which sat in its white bubble atop the mainmast in its new welded frame. The transceiver not only would allow Pawel to transmit video clips and interviews over the Internet but would give Watson dependable instantaneous e-mail and web surfing capability, which was much less expensive than having to use the dial-up modem on the Iridium satellite phone. ABC also gave Pawel $30,000 for his effort, which the Polish-speaking Australian promptly handed off to Watson to help purchase the new helicopter. The twenty-year-old chopper cost $150,000, and everyone knew it was cheap at the price because it might be the key to the campaign's success.

Aside from developing stealth re-breather equipment for the military, Pawel was the world's leading expert on the filming of sea dragons. A sea dragon is a fish that looks like leaves and branches. Or rather like a sea horse dressed up for a costume ball at which the theme is Your Favorite Flora to Hide In. Sea dragons blend in so well with the weeds and corals of the reef that Pawel has been two feet away, re-breathing like a fish himself, staring straight at one of the creatures, and unable to discern it. He will mark the waypoint on his wristband GPS and return six months later and the fish will not have moved more than ten feet. His films of the sea dragons are magical and beautiful, as are his shots of big rays and sharks. He wanted to dive in the ice with leopard seals, perhaps the fiercest predators Antarctica has to offer.

Pawel brought with him a tall, clean-cut, square-jawed, quiet assistant named George Evatt. They were both forty-two years old; their birthdays a day apart. George was also a prominent underwater film-maker, who shot a lot of Australian footage for National Geographic and Discovery. These guys could suit up and be diving in the ice floes in a matter of minutes.

Emily Hunter, the wholesome, ponytailed daughter of the recently deceased Greenpeace founder, while being on a bridge watch, was also on assignment for CityTV in Toronto. So she was shooting video and taking notes as well. There was almost noplace one could go on the ship without a camera whirring nearby.

The media on the ship had an interesting role. Since the media people were not on any of the watches, they had no assigned duties. You would not see a media person on the *Farley* swabbing the decks or cleaning the heads. Once in a while, Kristian or Mathieu would pitch in to chop a vegetable in the afternoon, but that was the extent of it. Unlike the rest of the crew, they were not under the direct command of any of the officers. While everybody else was working, they could be found in the mess working at their laptops, organizing their video clips or photos, writing notes, reading ragged Stephen King novels, napping, drinking coffee, bullshitting each other. Chris Price, the lumbering pilot of the FIB Thing, whose job was so specialized and so limited by weather he might get to fly only a few times, was usually there too, regaling the impressionable media with stories that usually involved whores, dead bodies, and a government conspiracy. He always started out, "Listen, wanna hear a story?" and ended with "True, I swear. He's in the phone book, you can look him up." Price used to go out with Allison, a long time ago, and he hung by the galley door while she did the dishes and serenaded her with conspiracy theories. About his job on board, he said, "I have another role on the ship, too, but it's secret."

And yet, like all embedded journalists everywhere, the media had to follow the rules in matters of safety, security, or general protocol. If the signboard outside the mess said, "No e-mail or phone calls today," that applied to the press, too. Sort of. And if the captain ordered a media person to do something, like wake up and put on a life jacket, or not to do something, like take pictures of the engine room, he had to do or not do that, too—sort of. There was a kind of a freewheeling attitude in the *Farley*'s press corps. During communications blackouts when Allison, who was the Robespierre of the ship, forbade all outside contacts, Taggart lurched out onto the aft deck, knelt, as if before an altar, propped his laptop on the row of bolted antique theater seats donated in Melbourne, fished out his personal Iridium satellite phone, waved it around for several minutes trying to get reception, and then conducted his commerce in e-mails. Some of the crew watched him with a touch of suspicion: could he be relaying the *Farley*'s position to an authority? Apprising someone of the ship's assault capabilities?

Pawel, too. He ignored blackout orders, went up to the radio room every day and surfed the web for news out of Japan, news of Greenpeace. He checked his personal e-mail. As the one technically in command of the ABC satellite Internet hookup, he had almost total access to the ship's computers, and 24/7 communication capability with the outside world.

It was not too outrageous to believe one of us was a government agent: in June 2004 the FBI declared ecoterrorism to be a greater threat to domestic security than Al Qaeda. This was in response to actions like the Earth Liberation Front's burning of the lodge on Vail Mountain in 1998. It was Allison's connections to ELF, and to the equally radical Animal Liberation Front (ALF), that brought her in front of a grand jury. Others on the ship also had connections to ALF. Given the political climate and Sea Shepherd's record of sinking ships and other acts of aggression on the high seas, and given the public announcements that Sea Shepherd was going down to Antarctica to physically stop the Japanese whaling fleet in Australian Antarctic Territory, I could think of at least three governments that might have been interested in having eyes on board.

Those were the five wheels of the *Farley*—bridge, engine, deck, galley, media. There were some specialists on the ship, like Marc the welder, Kristy the nurse, and Geert the ship's artist, along with the two pilots, who belonged to a category in themselves and reported only to the captain or his two mates. But it was all pretty loose and good-natured. Colin, the bosun's mate, could pass Geert and Marc on the aft deck and say, "Hey, can you give us a hand with the gangplank?" and they'd be glad to oblige. Usually they didn't have to be asked.

It costs a lot of money to keep a ship the size of the *Farley* afloat. Watson said that it cost $250,000 a month to run Greenpeace's *Esperanza* and *Arctic Sunrise*. Despite the generosity of his donors, Watson often didn't know if he had enough money to fuel up for a campaign until the tank trucks actually showed up at the dock and started pumping. On at least one occasion, the local fuel company simply gave him the tens of thousands of dollars worth of diesel with a smile and a "Forget it." Watson depended on the kindness of strangers and on the ingenuity and resourcefulness of his crew to keep the ship going. Nothing on board was wasted. In port, it was every crew member's responsibility to keep a lookout for anything the ship could use. When possible, the crew scrounged and foraged for spare parts.

The ship's bell had been rung by Saddam Hussein when it belonged to an Iraqi freighter. "They had a picture of it on board," Watson said. "The freighter had been seized during the first Gulf War. It was sitting in Bremerhaven, Germany. When we were there in 1997, we payed the Iraqi guard $6,000 and took off motors, a Detroit diesel, pumps. Got $150,000 worth of stuff."

When the *Farley*'s predecessor, *Sea Shepherd II*, was detained by the Canadian Coast Guard at Uclulet, Vancouver Island, in 1992 because Watson had refused to pay a $7,500 pilotage fee, he simply ordered his crew aboard and they stripped it to bare steel and recycled the parts. He then spoke to the local underwater diving association about donating the hulk and sinking it for a dive wreck. The Royal Canadian Mounted Police intercepted the phone call from Watson to the club president, and soon its agents were knocking on Watson's door, demanding, "What's this about you planning to sink a ship?"

After that first breakfast at sea I went up to the bridge to check in. The captain was on watch with his quartermasters Lincoln Shaw and Lamya Essemlali. Lamya, whom everybody called Mia, was a biology student from France who got into trouble with her university for refusing to dissect a mouse. Watson was playing his favorite comedian, George Carlin, on the bridge stereo. The sun was on the water and the mountains of Flinders Island were six miles off our port. Carlin was saying, "My dog Tippy was a 'mixed' terrier—you know, mixed, the word the vet puts on the form when he doesn't know what the fuck it is. Tippy was actually part Dodge Dart. Committed suicide . . ." Watson loved Carlin because he was an irreverent blasphemer. According to Carlin, the most ridiculous creations in the universe were human vanity and the institutions of religion and government. That's the way Watson saw it, too.

Watson was sitting in his high captain's chair, feet propped on the windowsill. I asked him what countries and groups still actively whaled.

He said, "The Inuit kill fifty-plus a year. In Alaska, Siberia. There's the Yupik on Saint Lawrence Island. Whaling nations are Japan, Norway, Denmark, United States (the Inuit), then Canada (Aboriginal), Iceland. There is a Russian hunt. The Faroese of Denmark kill 3,500 pilot whales. The Japanese hunt in the North Pacific in summer and Antarctic in their winter. We could go after them in the North Pacific but it's even harder out there."

Watson talks in bursts, in a persuasive, confiding rush of fully formed paragraphs.

"Our biggest problem over the years," he continued, "is how to control political correctness. Sometimes it gets out of hand. Lamya, who is French, was calling Marc and Alex racist because they said they don't date Dutch girls. I had a Newfoundlander call me a racist the other day. The Makah Indians kill whales. If you're antiwhaling, you are anti-Makah, you are a racist. The one I like is, 'You are a privileged white North American male.' I was raised in a poor family with a bunch of kids.

"Anyway, the Native American gets a free pass on environmental destruction. People don't realize that by 1820 the buffalo herds had been reduced by 75 percent. There were roughly 3 million horses on the plains by 1820. The horse was displacing the buffalo, and the Indians went after pregnant cows. Estimated they took out 4 million to 5 million buffalo a year. By 1860 the great herds were almost gone. By the time of the white hunters. Both Indians and white hunters worked in partnership.

"Back to the Makah of Neah Bay. It means, 'People of the Cape.' They used to bury homosexuals up to their neck at low tide. This was their punishment. I was on the radio with a Makah; he said, 'We lived here from time immemorial, we are the people of the cape.' I said, 'Yeah, well, don't give me that stuff—you came from Vancouver Island, you massacred the Ozettes, stole their land.' Now the Makah said, 'Oh, the Ozettes are our ancestors.'

"At the IWC meeting in Monaco in 1997 we had a couple of elders in our camp. The Makah were petitioning to kill their whale. Tom Mexis Happynook, head of World Council of Whalers . . . was raised in a suburb of Victoria. He said, 'My people go out on the waters to hunt the great whale. Our brother the whale gives himself so that we might live. The women go out on the beach to cut up the whale for several days, so that our people might live.'

"After twenty minutes of this shit, Simon Dick of the Kwakiutl shows up and says, 'My people were never that hungry.'

"In 1996 when we went into Neah Bay in the *Edward Abbey* [Sea

Shepherd's converted Canadian patrol boat, which had since been donated to the rangers of the Galápagos], we met a canoe full of whalers in feathers. The Makah! In feathers! They never wore feathers.

"They said: 'White man, you come with evil in your heart to interfere with us killing a whale that the people might live.' The news cameras were rolling. I said, 'Spencer, you sell most of that whale meat to Japan.'

"'When our brother the whale gives himself up so we might live it is a something between me and my brother.'

"The television lady said, 'Your shaman said—'

"I interrupted. 'What shaman? Everybody is Christian.' On the debate, I said, 'I'm not up here to debate a native whaler, I'm debating a Japanese whaler.'"

Watson was getting angry as he remembered. He tilted his head forward and looked out like a bull thinking about charging.

"At talks when I'm speaking, I hate it when they do this: the stare.

"They say, 'I am one of the two-legged ones. You have evil in your heart.' They must get all this from the same script. They stare. Minutes pass.

" 'What's your question?'

"People shout. 'Let him speak! He's Native American!'

"Ten minutes later I say, 'Look, I've got to catch a plane.'

" 'You're not going to hear him out?'

" 'I'll hear him out, but he's not saying anything!'"

Watson shook his shaggy hair out of his eyes. "I've seen Inuit on the Saint Lawrence Islands shooting walrus with M-16s for their tusks."

Watson swung his feet to the deck and went into the chart room. Carlin cut off and a second later a monumental string section took his place. It was Wagner's *Ride of the Valkyrie*. It shook the bridge and blasted from the outside speakers over the decks—the same music that thundered from the squads of attack helicopters in *Apocalypse Now*.

7

Lifeboat Drill

By midday the coast of Tasmania came into view off starboard, low flowing mountains covered with forest—eucalyptus, pine, gum trees—falling to cliffs and the sea. A few farms carved out of the slopes above the beaches. The ship alarm sounded, and the crew scrambled for the aft deck, and we had our lifeboat drill.

It was brief.

The second officer, Hammarstedt, told us that if the boat sinks it will be most likely from a hull breach caused by an iceberg or another ship. He said the ship will take from twenty seconds to eight hours to sink.

He pointed to three lifeboat pods on port, starboard and stern. They were white fiberglass barrels, about half the size of a steel drum, nested on cradles by the rails. Each would be commanded by one of the ship's top three officers: Watson, Cornelissen and himself. He passed around the roster for the boats, and I noticed that I was with Watson. In fact, all the media were with the captain. In case of a sinking and lifeboats adrift, I guess he wanted to give daily press releases.

Hammarstedt said each raft was equipped with a knife, flashlight, enough water for about seven days. He said you just unclip the lashings and two hands throw it overboard, and it will explode open. "Put on as many clothes as possible and grab water from the galley if you're there. Don't jump into the life rafts. Do not. They will fill up with water."

That means the quick way in was to jump overboard into the drink and then climb into the raft. And the rafts? Given how the Zodiacs had functioned in Melbourne harbor, and how the Jet Skis did not function at all, I wasn't giving life raft deployment the best of odds. I could imagine tossing them over the rail and seeing an explosion of confetti and balloons and a banner that said "Surprise!" Then Hammarstedt pointed to three waist-high black lockers on the deck. "The immersion suits are in this locker, and the life jackets are in these two. There are a bunch of immersion suits. The rest will wear these Mustang suits, which are in the locker in the hallway outside the mess."

The life jackets were mostly bulky Mae Wests that felt as if they were filled with horsehair and newspapers. The immersion suits were orange neoprene coveralls with attached hoods, gloves, and booties. They were stiff with age. Hammarstedt suggested we try to put one on. They did not look as though they would seal very well around the face, and if they did not seal, they would fill with water. Little Kristy, the hippie nurse, stepped into the smallest one and disappeared inside it; her hood flopped empty like the Headless Survivor, only a few strands of dreadlock peaking out. Alex said, "We'll try and find you a wet suit."

The mustang suits were insulated nylon coveralls with no water seals at the wrists, ankles, or neck. They would fill with water immediately, and without latex closures the icy current would wash through them. They would be about as good as wearing a snowsuit in the thirty-degree antarctic water. I was very glad I had my own dry suit. I decided that when we reached Hobart, I would buy my own life vest and a thick neoprene hood. If I was going to be in the water, I wanted some time to think about things.

"Oh," said Hammarstedt, "Don't piss off the stern. Seventy percent of the bodies they've found at sea had their flies undone." Man. Think of it. You make your way to the stern to take a leak under the stars, the ship lurches, in you go, treading water and yelling as the running lights diminish. "Even if someone saw you," he added, "it takes fifteen minutes to turn the ship around. You have less than two, probably less than one. So if you go overboard, you're dead."

As the crew broke up, Ron approached Laura, who was standing on deck in her flip-flops, hugging herself. He held out a pair of sneakers. "You'll need these where we're going," he said. She brightened, then asked, "Are they vegan?"

That evening, I climbed to the crow's nest on its mast twenty feet off the top of the superstructure and watched the sun set over the hills of Tasmania. It set fire to the forests of the ridgetops which the smooth sea reflected and the cool dusk put out. A pale violet ringed the horizon just over the sea, scaling into luminous blues that got deeper as they climbed. A cry came up from the high forecastle at the bow. A mile off, a white thrashing on the port side. Off starboard, between us and the coast, another commotion. The two disturbances were closing, and then we saw the blue forms leaping. The fins, torpedo-fast, and the arching jumps. Two pods of dolphins, running hell for breakfast to meet the ship. How far had they come? The ship's whale alarm sounded. I clambered down and across the main deck and up to the bow and squeezed in with a bunch of others to hang over the rail. They ran under the bow, turning right angles with no drop in speed, with the fluidity of water, catching the bow wake with no effort at all, ranging out in a weaving V of escort, skimming the surface and turning it to a rippled glass that magnified what seemed less form than pure blue motion. Then broke, as they leaped each other for the post and canted half over to cast one curious eye upward at the heads hanging over the rails. They blew in hollow gusts, a staggered rhythm, and kept up a commentary of light squeaks. Often they seemed not to move a muscle but maintained the nine knots by some contract with the sea, as if their whole being were a single thought: watergray and swift; and a single emotion: breaking white glee. They had come to the ship like two squads of kids running to a big game of kick the can, which is exactly what we were.

Recently scientists have proved that dolphins can not only call each other by specific names, but in communication can refer to a third animal by name. They can form shifting alliances—a very sophisticated form of social interaction—and though they lack the part of the brain that governs self-awareness in humans, in controlled experiments with mirrors and body paint they prove to be extremely self-aware.

They seem fantastically superior beings: they don't spend too much time working for a living, they are social and expressive, they have a highly developed sense of fun—stories abound of dolphins sharing waves with surfers—and they don't trash their nest or decimate other species.

The Japanese issue permits to kill over 22,000 dolphins, porpoises, and other small cetaceans every year. In Taiji, in Wakayama Prefecture, they corral the dolphins into a netted-off bay and club them, slit their throats, and spear them. That the dolphins exhibit human-like traits of loyalty, valor, and grief does not affect the fishermen; the dolphin being killed often cries out in front of his family members, who swim frantically around trying to defend him.

It is this killing that Allison Lance and most of the crew of the *Farley* considered murder, that they could not abide. Most of them, in fact, thought the human species was a terrible aberration on the tree of life that had turned on the earth and was senselessly devouring and destroying her. They thought of the human race as a cancer.

It was no wonder, then, that many on the *Farley* wanted to take up as little space on earth as possible. The prevailing view on the ship was that having children—"plopping out babies" was the favored expression—was a crime. Indeed, it was a sin. Knowing what one knew about what humans were doing to the innocents, about the impact of a single person born into a high-consumption developed economy, to have a child was blasphemous. Since one was already here and had no choice in the matter, the most ethical path for the greatest good was to be an activist, to mitigate the harm other humans were causing, and to walk as lightly as possible. Gedden, for instance, was a "freegan" who, at home in Cascadia, endeavored to get most of his diet out of dumpsters. Since he was eating food that was going to be thrown away, since he scavenged for clothes and just about everything else, he was consumer-neutral, or as close to that as one could be in today's world. (Paradoxically, this meant that he ate meat, which irked his girlfriend Inde no end. Sectarian strife was rife among the vegans, freegans, and vegetarians.) There was a bit, or a lot, of self-abnegation in this way of being—even self-hatred. There seemed to be as much about fanatical denial in all these isms as there was about saving the planet. When Kristian reached for the little honey bear the first night at sea, Laura said, "Did you know honey is cruel?" Startled, the videographer said, "Why?" "Because they make the bees slaves and then they steal the honey—"

I asked Allison whether, if she had a button in front of her that would painlessly evaporate all the humans off the face of the earth, she would push it. She said without hesitation that she would. She hoped some epidemic like AIDS or bird flu would wipe most of us out and bring the population back into balance. The savvy captain, on the other hand, wouldn't give a straight answer.

"The planet's chances of survival without us are greater," he said. "Would an elephant press that button? Would a dolphin? Probably. Just out of sheer self-defense."

"And you?"

"We don't hate anybody," he said. "We look upon humanity as a problem as a species. So-called misanthropes tend to have more respect for people as individuals. I could press that button in the abstraction, but I have a lot of respect for the people doing this. Consider that there are 6.5 billion people on the planet, and you can count the number of activists on two hands."

8

Hobart

December 12, 2005

The next morning at nine we rounded the high cliffs of Cathedral Rock guarding the eastern cape and turned north into Storm Bay. Since dawn the south coast had been a ragged rampart of tall fluted cliffs and sharp guard rocks at the mouths of rugged coves. Low scudding clouds and damp air. Fog boiling over the tops of the headlands. As soon as we turned the corner, the wind hit, twenty knots offshore from the northwest, and cold, raking the bay into gray chop. Hobart sat at the top of it. Watson said we'd be there by noon. The chopper needed a part that was being flown in on a six p.m. flight, so we wouldn't leave until 1900 or so. Seven hours in port.

A lot of people had called Watson an ecoterrorist, including several governments and, in a fit of pique, Greenpeace. One group, the Center for Consumer Freedom, had said on its website that the bow of Watson's ship was reinforced with concrete for ramming and had a steel blade called the "can opener" for gutting the hulls of ships. It claimed that he had AK-47s on board. I thought I'd better ask him about all that. I climbed up to the bridge and found the captain on watch in his chair. He was reading *A Devil's Chaplain*, a book about evolution by Richard Dawkins. He was playing Jimmy Buffett on the stereo, to the horror of everybody under forty.

He closed his book. "I don't know of any ecoterrorists other than Union Carbide and Exxon," he said. He meant the environmental catastrophes inflicted by the *Valdez*, and the chemical plant explosion at Bhopal.

"You heard what the FBI said about the domestic threat of ecoterrorists?"

"We're not identified as an ecoterrorist group by the FBI."

He said that a coast guard officer had given a talk in Port Angeles, Washington, and described Sea Shepherd as an ecoterrorist organization. Watson wrote a letter to the commander of the coast guard in Seattle and got an apology from him; the officer was reprimanded. He officially asked the FBI if there were any investigations into Sea Shepherd and was told no.

"We got audited by the IRS—didn't lose our 501C3 status for our activities, so—they don't give nonprofit status to terrorist organizations. If we were terrorists or pirates we couldn't dock in the United States. The trouble is, every right-wing antienvironmental group in the country runs around calling everybody an environmental terrorist and everything." He added that nobody was doing what Sea Shepherd was doing. "We're filling up a niche. Intervening against illegal activities. We've been called sea cops, eco vigilantes."

When asked if he thought he was putting out little fires in front of a massive conflagration, he said, "I'm sure that the whale that we save isn't too concerned with being a little fire."

"Is your bow reinforced with concrete? Do you have a can opener?"

"The bow of the ship is not reinforced with concrete. We had a can opener. It was on there once, but we took it off; it was getting rusty."

About guns he said, "Well, we use shotguns to destroy the buoys on longlines. Beats jumping into the water with an ice pick."

He said there were no AK-47s on board, and that he'd never had to use any weapons for self-defense. "They can say whatever they want, but we don't shoot at people."

"Cannons filled with pie filling," Allison piped in. She was sitting in one of three wooden seats mounted on the bulkhead just behind Watson's chair.

I had wondered about the three-inch cannon lashed on the main deck.

"We fired pie filling," Watson said. "That was the Faroese. When they tried to board us, we hit them with forty-five-gallon shots of custard and banana crème."

Since I was airing all my questions about terrorist tactics—my concerns, really—I asked, "The action that resulted in, say, the sinking of the Icelandic whalers—how do you make sure nobody's on it? That you don't kill anybody?"

"Search the boat. The watchman, he was sleeping up on the bridge. They cut that boat loose. That boat drifted away from the other two. He was in the middle of the harbor when the other boats sank. He was still sleeping."

"Searching one boat in Norway," he continued, "there was nobody on it, it was total dark; went in the galley and some drunk Norwegian was lying on the mess table, passed out. He was so drunk they could have picked him up and put him on the dock." Instead, they left the boat alone.

Watson added, "People call us pirates, but we don't go robbing anybody, so we're not pirates. Kids just like the romance of it so we do the T-shirts and stuff. We simply don't break laws. The perception is there that we do. If we were pirates we couldn't come into ports like this."

Sink ships but don't break laws. He said again and again that he was simply destroying property used in criminal activities, upholding international law.

At the chart table, Alex was drawing up a design for a steel frame to hold a wood barrier to protect the barrels of aviation fuel under the heli deck. He said if there was a crash, he didn't want pieces of rotor blade slicing down into the gas depot and blowing up the ship.

The chart room was, for ship's space, a big open cabin. The chart table with logbook and GPS display was on the port side. At any time, one could look up there and read the ship's position, speed, and heading. On the starboard side was a long varnished counter that held the computer monitor that displayed the RayTech navigation system, a sophisticated program that allowed the officers to display the ship's position in relation to any number of waypoints anyplace in the world. Cabinets stuffed with signal flags, flares, smoke bombs, and megaphones ran under the counter.

The rest of the chart room was lined with bookshelves, each with its strip of guarding wood to prevent the volumes from going airborne, so the books seemed to be looking over the top pole of a corral. One small case held marine animal guidebooks: fish, whales, bird guides. Another, over the chart table, held Admiralty Coast Pilots, with detailed information about tides, channels, buoys, hazards, lighthouses, and ports for every coast in the world. The one we would be using a lot, *Admiralty Sailing Directions—Antarctica Pilot*, had a lot to say about ice. The bookshelf over the nav computer held the most curious collection of titles:

Reserva Marina de Galápagos
U.S. Navy Seal Combat Manual
Criminal Evidence, by Klotter
Navigation Rules
Modern Criminal Law, by La Fave
National Legislation and Treaties Relating to the Law of the Sea—
    United Nations Legislative Series
International Environmental Law—Basic Instruments and References,
    by Weiss, Stasz, and Magraw
International Environmental Law, 1994 Supplement, by Kiss and
    Shelton
Black's Law Dictionary

Moving from left to right, from the *Navy Seal Combat Manual* to *Black's Law Dictionary*, the titles created a narrative that in some way elucidated Watson's approach to direct action and helped explain why he wasn't behind bars right now. The books went from combat to criminal evidence and on to the broader dictates of international treaties, which, evidently, had checked any local prosecutor's enthusiasm for indicting Watson. A state like Norway simply did not want to focus world attention on its whaling program by giving someone like Watson a public forum in court. Such states were already flying in the face of world opinion and international laws. Putting Watson on trial would be bad PR. Watson enthusiastically exploited this dynamic. After he had sunk the three Norwegian whalers, he repeatedly and publicly offered to return to Norway to stand trial. He said he'd pay for the ticket. After he rammed two illegal Japanese drift netters in the North Pacific in 1990, the Japanese refused to admit that the incident had happened.

In the case of the Japanese whaling fleet in Antarctica, circumstances were a bit different. The Japanese didn't seem to care much about PR in this arena, aside from publishing booklets that justified from every angle the need to kill whales in great numbers. One full-page chart put out by the Japanese Institute for Cetacean Research showed how whales were depleting the world's fish stocks.

They were certainly not keeping their activities quiet, and they were moving boldly in the IWC to lift the moratorium on commercial whaling completely. They'd been trying for eighteen years.

So why wouldn't the Japanese captain of the fleet just blow us out of the water if Watson tried to ram him? Or seize the ship, and take everybody back to Japan, and lock them—us—up?

I guessed that the answer was that there are degrees of bad international PR.

The chart room—being the locus of much decision making and current information on the ship's whereabouts; being spacious and full of books; having counters on which to drop camera gear and of a perfect height for leaning against; and, not the least, being between the bridge and the radio room—was a natural gathering place. It also held the only hatch from inside the ship onto the sauna and heli deck. Anybody on board who wanted to see where her efforts were taking us, or snatch a whiff of the latest plan, would climb the steep, narrow stairs and hang out in the chart room. From the bridge, as you moved aft across the chart room, in fifteen feet you had a choice: turn right down the stairway; continue straight into the little radio room; or move left, out the main hatch which led to the heli deck. The radio room was a cramped office holding an L-shaped desk, a bench seat big enough for two at a squeeze, and a padded office swivel chair which put the sitter in reach of everything in the cabin. The two satellite phones were in here, as well as the ship's main computer, and a marine radio. It was also the ship newsroom, press office, and PR Office. From this creaking seat Watson conducted interviews with news agencies around the world, and churned out, night and day, a steady stream of press releases, op-ed pieces, e-mails, historical essays, and diatribes; and from here he published, on a sporadic basis, the ship's newsletter, the *Scuttlebutt*. The radio room had a steel door with a bolt and a reinforced steel bracket for holding a stout bar. Its final function was a safe room in case of boarding. From here, Watson could transmit distress signals or fire off a last volley to the press before being seized.

The harbormaster in Hobart had, unfortunately, read the Sea Shepherd website. He mentioned it over the radio. In Melbourne, the port authorities had waived thousands of dollars in docking and piloting fees under a provision for the ships of charitable nonprofit organizations. Watson had paid a pilot fee coming in, but no pilot fee going out, and no port fees at all for the nearly three weeks tied up downtown. Captain Mike Bass-Walker didn't see it the same way and would not give Sea Shepherd a freebie. The cost for a pilot and a berth in the dock for a day would be over $2,000, so Watson took his anchorage coordinates in the free zone three miles from the wharves, and at 1230 the deck crew up on the forecastle head lowered the anchors.

They had meant to lower only one, but in lifting the steel block off the windlass they had released both chains. Alex screamed out the window from the bridge. "Stop the windlass! Both your chains are going out! Stop the windlass!"

Kalifi stood in the stiff wind and the clatter, her hair blowing wildly, and looked clueless. The other hands swarmed around, not sure what to do.

"Jesus." Alex shot out to the bridge wing and down the outside steps, and ran up to the forecastle and took charge. The port anchor grabbed and the *Farley* swung up into the wind and the chop. Watson was never embarrassed by the sort of cluster fuck exhibited on the forecastle. When he still sailed with Greenpeace back in 1976, he had a Czech captain who used to splinter the pier with every berthing. He said they'd rate the dockings One, Two, or Three Yells. Watson thought it was funny. He could be on the bridge with an entire harbor watching him smash the wharf and not cringe. With an all-volunteer crew he had to be pretty much immune to that sort of thing.

The boat rode at anchor in the short chop with the undulant motion of a cantering horse. Nobody was allowed onshore until customs had cleared us, so the Jet Skis were immediately plucked from their cradles by the port davit and lowered into the windy bay. Geert and Gedden started them up and began roaring around the ship. Practice for attacking the whalers. Gedden was a wild man. He did doughnuts, aerials. He leaped a wave and flipped. By the time Geert could tow him to the ship, the ski was barely floating. He'd almost sunk it. The engine, which had just been tuned, was swamped with salt water. Gedden had wrecked the one on Henderson Island as well. Then customs cleared us, and two Zodiacs bounded into the docks to pick up supplies. One mission was to get Allison a pair of goggles so that she could ride shotgun with Price on the back of the FIB thing.

One Zodiac was loaded in Hobart with almost nothing but booze. Cases and cases of beer and bottles of rum and whiskey. On the way back out it swamped in the whitecaps. The other Zodiac stalled in the middle of the bay with a blocked fuel line and wallowed dangerously in the swells. In the confusion, someone dropped one of the three working handheld VHF radios overboard, and the other got soaked.

By the time the liquor boat got near the *Farley*'s side it was full of water, "bow up like the *Titanic*," Watson said. Wessel started chucking beer and liquor overboard in a frantic effort to save it.

Watson boomed out the bridge window: "Quit throwing the booze in the bay, will ya?"

The waterlogged Zodiac was secured, and it was so heavy that it bent a cleat on the ship's rail, and the line caught and wrenched Colin's hand. So now the bosun's mate was partially dismasted.

In the mess, Jeff of the J. Crew was drying off and drinking a cup of tea. He mentioned that one of the radios had been lost. Watson happened to be standing behind him and stepped forward, looming.

"It's *lost*? Who lost it?"

Jeff put his head down, the former inmate of solitary balking at the Man. "Aw, c'mon."

Watson, steely, overbearing, not messing around: "Who lost it? Tell me."

"Aw, c'mon Captain, you want me to rat?"

"Tell me now."

Direct order from the captain.

"It was—" he mouthed the word. The captain stormed out.

Jeff put his head down on his arms. "Aw, fuck. I'm a straight-up bitch now. I'll be washing socks and sucking dick for the rest of the trip."

There was a big note scrawled on the message board outside the mess:

Only officers may use the handheld marine radios! Do not take one without asking. We've lost half of our radios in one day and we haven't even gotten to Antarctica!

On the bridge, Hammarstedt's watch was somber. Too many mess-ups in one day. Hammarstedt didn't seem too discouraged. "This always happens," he said, "at the start of a campaign."

A few hours later, the flight deck crew had swung down the shiny new rails around the heli deck and readied the cam-strap tie-downs for the arrival of the chopper. The new rubber was down on the deck and painted with a white center stripe to help Aultman line up. Aultman had handpicked Lincoln, Dennis, Gedden, and Colin to handle the deck and the chopper during takeoffs and landings; but now that Colin was injured, he was replaced by the tall, multilingual Casson. Aultman had trained his crew to guide him in with hand signals, and as soon as he touched down, their job was to rush forward and hook the pontoons to the newly welded rings before the chopper could think about sliding off the ship and into the sea. The rings were the cutout circles Marc had made in the deck, with rebar rods welded across them from underneath, so they could be hooked to the deck but not project anywhere above its surface.

It was nearly eight p.m. (2000) and the helicopter part had arrived and been installed. The chopper had been cleared by the mechanics, and though the marine band radio didn't work, they'd have to make do. The previous owner, who had also flown it off the back of an excursion boat in Antarctica, radioed the *Farley* to say that Aultman had taken off and would be arriving within fifteen minutes.

The fire crew took their stations, and all decks aft of the superstructure were cleared. Aultman and his crew on the heli deck would be executing their first landing at sea, a truly dangerous maneuver. The way things had gone today with all the other *Farley* auxiliary craft, everybody was a bit tense.

You might have thought Aultman was Lindbergh landing at Paris. Fire hoses snaked aft, with a hand ready at the valve and another training the nozzle on the barrels of fuel and on the heli deck. The entire crew lined the rails. Even Watson lumbered out from the radio room where he'd been composing a press release blasting the personal character of the director of the Institute of Cetacean Research. He said that when Hiroshi Hatanaka was a student he had bragged about cutting up dogs alive. Soon a fast-moving speck appeared over the harbor, and grew, and then we heard the signature throp carried downwind and saw the neat red-and-white whirlybird. It was small, not quite toy-like—a Plexiglas bubble; a thin spine of tail, engine, pontoons; period.

It seemed diminutive for the vast, raging landscape of Antarctica, but on the upside, it looked light enough to float for a while.

Aultman buzzed the ship to cheers and circled it twice. He came in over the stern, over the heli deck, and hovered, gave the pitching landing pad a look, and then lifted off and circled again. He came in a second time and brought the chopper within feet of the black rubber, yawing slightly, trying to center the white stripe, the stiff wind and pitch of the boat not helping. We could see him now in his drab flight suit and helmet, right hand on the stick, continually adjusting it, completely focused. Abruptly he lifted, accelerated, tilted forward, and took a wide sweep of the bay. A third time he came in. Lincoln centered him with straight arms. Aultman brought the chopper down to within inches of the tossing rubber, levitated, then dropped it down. The slack pontoons gave, pancaked, and rebounded. The chopper stuck. The flight crew was on it in seconds, running on the crouch, securing the machine. The ship cheered. The wind whipped. Three seagulls flew by wondering what all the fuss was about. Aultman had never landed on a ship before, and after all the fuck-ups it was heartening to see something so beautifully executed.

Within a couple of hours we weighed anchor. The *Farley* swung off the wind and with dusk drawn down on Storm Bay melted into the darkness. The lights of the last land slipped astern and flickered out. I checked my GPS. We were heading 176 degrees, four degrees east of due south. Watson had told the media and the crew in Melbourne that we would head south and west, toward the Greenpeace ships coming from Africa, to squeeze the whalers in a kind of pincer action. He had told me we would head more eastward. Now we were heading almost due south and I wondered where we were going.

We weren't an hour out of Hobart. Our life at sea, beyond the reach of port authorities or police, had truly just begun.

"You in, dude?" Justin of the J. Crew shuffled the cards at a booth table in the mess. He shuffled them flat to the table with the tight riffle of two corners, like a pro. He had long dark eyelashes, a black beard, big glistening black eyes. He looked innocent, benign. And he could wipe you out of a paycheck in twenty minutes. In first grade he had won a kid's cat shooting dice.

"Uh, no," I said, following Price with my eyes.

The FIB pilot with the secret second job was carrying around the bolt of a high-powered rifle. He stopped by the galley counter and wiped it down lovingly, then went around the corner into the main companionway. My heart raced. There was no mistaking the smooth blued steel, the weight of it in his hands.

I was pissed. It was not so much the gun itself, the piece of it, that shook me. I'd been around guns my whole life. What prompted the wave of anger was betrayal. And fear. Watson had manipulated me baldly. He had violated my trust. To me he was an unknown quantity with a wild and dangerous reputation, and despite his bearish charm I knew him to be a man of iron will who held the laws of men in contempt, especially when they interfered with his protection of wildlife that had no advocates but himself. I was suddenly afraid that Watson had a deadly agenda of which few others on the ship were informed. We were all in this together. The thing about a ship like this is that there is no getting off.

The day before he had talked about his near ramming of a Cuban bottom trawler off the Grand Banks, for which he was arrested and charged by the Canadians with two counts of life threatening mischief. He had faced a possible sentence of two life terms plus ten years. He said, "I was the only one responsible, and the only one charged. It's one of the reasons I don't have crew meetings—to protect the crew from charges of conspiracy."

Price ducked back into the mess and continued on into the green room, where he sprawled on a cushioned bench seat. I followed him.

"Hey, Chris."

"Hey."

"You want tea water?" I grabbed the electric hot water pot.

"I'm OK. Thanks."

"Nice bolt you were carrying around. Looked pretty high-caliber."

Price always seemed a little amused about everything, even his own fate. Must have come from circling the earth for days on end under a few square yards of gossamer nylon. He pushed his steel-framed glasses up on his nose and looked away like a kid who had stolen a candy bar. "I don't know anything about that," he said.

"I don't give a shit what it was," I said. "I'm a hunter. Hunt elk every year I can. Savage 99, .308. You know, the one with the safety on the tang?"

The mention of a specific gun—make and model, caliber—was too much for the big trucker from Arkansas. There was not a soul on this ship full of vegan hippies he could talk guns with, except maybe DMZ Steve, who wasn't much into idle chat. It was like hitting the On switch.

"That's a lot bigger than a .308," he said. "It's a .50-caliber BMG."

"Oh, shit?" I said, making to head for the clean water jugs with the pot.

"Two thousand pounds of pressure. A guy just set the world record on five shots at 1,000 yards. A 2¾-inch group! Think about it!" He was beside himself. I turned.

"Could blow right through about anything."

"Right through the engine block of a Humvee!"

"Through the hull of a ship, even."

"Piece of cake!"

"Better keep it oiled good out here."

"Yeah, the guns got wet in the crossing."

Guns. How many frigging guns? Fifty-caliber. Serious firepower. What the hell was going on? I took some deep breaths, trying to control my fury and my fear.

The captain's cabin was a sanctum. I lifted the brass whale's tail and hammered.

"Come in!" I turned the ornate handle and pushed, and wasn't at all prepared for the genteel scene within. I had lived for the past week with the cramped quarters of an old trawler, and the spaciousness of the cabin struck me as luxurious in the extreme. It was as big as the bridge and half the chart room above it, and appointed as well as any ship's cabin I'd ever seen. Dark wood paneling, portholes forward, two sumptuous settees facing each other across a low coffee table. Bookshelves, a chart table, a double bed with headboard and fluffy duvet. A private bath with a Jacuzzi hot tub. Watson later told me that Paul DeJoria of Paul Mitchell hair products had commissioned the cabin for the Watsons as a gift, soup to nuts: $75,000. Watson spent much of his year on the ship and he seemed to work twenty hours a day. If some benefactor wanted to provide decent living quarters, bully for him. It was like stepping out of the Astroturf carpeting and fluorescent lighting of a Days Inn hallway and into a room at the Ritz.

The captain and Allison sat across from each other on the love seats, each holding a drink. "Oh, hey," they chimed. "Come on in. Want a drink?"

They took a look at me and their demeanor changed.

Allison, crisp, no longer pleasant: "What's up?"

"How many high-caliber rifles do you have on board?"

Silence.

"I just saw the bolt of a .50-caliber BMG. That's a sniper rifle."

"No, it's not," Watson said.

"Yes, it is. It fires a 2¾-inch group at a 1,000 yards." I'd never seen a .50-caliber, but anything that accurate at that distance was a sniper rifle. It was certainly not designed for self-defense.

"It doesn't fire in a group. It's a single-shot recoilless rifle."

Watson sat on his settee, drink in one hand. His voice was calm, relaxed, with the almost offhand, confiding tone he used whenever he was mining the narrative field with deadly facts, as he did all day long. He lived in a state of war and acted as if he were sitting on a picnic blanket, handing out fried chicken on little napkins. Allison, by contrast, sat on the edge of the love seat. Her natural tautness was now the stillness of a big cat about to leap. Her mouth was tight and her eyes blazed.

"No," I said to the captain, "I mean you fire one shot, reload, fire another shot. Five shots making a group like this."

"We've had it onboard for four years. We've never used it. It's missing a part. I don't even know if it works."

Price had a secret role and he was lovingly oiling the bolt to a junk rifle.

"How many guns do you have?" I said.

Allison cut in. She was leaning forward as if she might spring off the settee. "You're a hunter. Why are you upset about guns? You should understand!"

Watson held up one pacifying palm toward his warrior wife.

"We have two shotguns and this rifle," he said.

I sat against the foot of the double bed.

"Is it part of your plans for the campaign?" Would he tell me if it was?

"No," he said without hesitation.

"You've never used it? It's backup?"

The captain took a sip, put his glass down on the coffee table. "Sometimes, in some places you want a gun. We sail in places where there are real pirates."

I nodded.

Allison glared. "Every boat has weapons on board."

Our eyes locked. We had a long trip ahead of us. I looked back at the captain.

"OK," I said. I got up and walked out. My fears about Sea Shepherd seemed to be bearing out. To what lengths would they go to protect the whales?

9

Southern Ocean

December 13, 2005

43°33'S, 147°20'E

I woke at three a.m. braced against the rails and was nearly thrown from my bed. The ship groaned and lay over, paused, rolled back hard. I groped for the light and tensed and timed my launch out of the bunk so that I wouldn't be fighting the G forces of the roll. I slipped on a jacket and clambered to the bridge, gripping the handrails. The bridge was dark but for the pulsing green of the radars and the glow of the compass. The sea was glorious. The moon scudded in a roiling overcast and was gone. The ocean reflected the light like a smoked mirror still molten and flowing. It heaved darkly, gleaming back the moon from the tops of the swells, then dimming to a vast plain of black motion that seemed to suck the umber from the edges of the sky and left a luminous ring at the horizon. An endless procession of steep waves rolled out of the northwest.

Hammerstedt said we had just cleared the southern cape of Bruny Island, the last sheltering land. "That's what you're feeling," he said.

"It's open ocean now all the way to Antarctica. It's all downhill from here."

I stood on the dark bridge, braced into the port corner. The cape rollers barreled from astern on to the starboard quarter, so the ship rolled, pitched, and yawed, moving in every direction it could at once, and sawing back. I tried planting my feet and swaying in a fluid circle, leaning into the tilt of the deck—one wave, two—and lost my balance. I crashed halfway across the bridge and stout Gunter caught me.

I waited until the ship canted against me, and pushed through the heavy hatch on the port side. I didn't want to do it on the downslope and go flying through the door and over the rail. The cold air washed over my face. It had been chilled by miles of open ocean. Here was a little covered deck, no bigger than a bathtub, with waist-high bulwarks and a rope safety railing. It was called the bridge wing. It was a great perch, as it sat just behind the bridge, one deck on each side, so you could take the one on the lee and stand outside but sheltered, watching the weather and spray blow by.

I let the door slam closed with the next pitch. For a while I just watched the dark sea and listened to the hiss and rush as the bow labored out of the troughs. Stairs went straight down to the poop deck. I gripped the rope banister and went down the steps. I was on the port side, a few feet from the railing. At the bottom I held tight until the ship rolled to starboard, away from the rail and the cold ocean, and ran as best I could aft. The aft or poop deck was partially covered by the heli deck. I ran as far as the square lockers holding the life vests and grabbed a support pole. Steadied and made another burst to the next pole. Someone had lashed a little glass LCD lantern there for a running light. They'd backed it with a piece of cardboard that flapped in the wind with a snapping flutter. Then I made a last lurch to the stern chains—two steel chains strung across the high open stern. There was a place on either side, the rear corners of the ship, where you could wedge against the turn of the bulwark before it gave way to the swinging chains. With one hand on the top chain and one on the pole supporting a running light at the corner, you could look out over the ship's wake in the roughest seas.

The wake had a phosphorescent glow, trailing into the hills of water. Even in this darkness, I could make out the pale forms of birds, probably shearwaters, swooping over the white road, soaring past the stern to circle the bow. Silent ghosts.

I made my way back to my cabin and fell asleep until breakfast.

1.4 million square miles was too big. Watson said he had all these sources of intel, but so what? Having eyes at a few research stations and on a few supply flights was not enough. After breakfast I asked him if I could e-mail my friends at SpaceImaging, a private satellite imaging company based near Denver. They had proved invaluable in obtaining images for the last expedition I'd been on, in the Tsangpo Gorge in Tibet. Watson said, "Sure."

I wrote from the radio room, "Can you possibly look for a group of four to six sixty-meter to 130-meter ships between 30 degrees east and 175 degrees east within 200 miles of the ice?"

The response was almost immediate. My friends said they'd need approximately 800 shots and 100 days to scan that area. They said it was like looking for a needle in 100 haystacks.

A needle in 100 haystacks with eyes in space. What did that mean from the water?

Despite the discouraging information, Watson was energetic on the bridge. He tortured his watch with Canadian folk music.

"This is supposed to be the Roaring Forties. The Furious Fifties. The Savage Sixties. The Shit Storm Seventies." He was talking about degrees of latitude. A degree of latitude consists of sixty minutes, each representing one nautical mile; so with every ten degrees we went south, we were 600 miles farther from an inhabited country. "We didn't have a single good day crossing the South Pacific. We're pretty heavy right now. Rolling a lot more when you have less fuel."

A steep roller lifted the ship and barreled underneath. Half of us were thrown across the bridge. The *Farley* rolled back just as hard. The little roll gauge over the center window swung out past thirty degrees. Watson got up from his captain's chair and walked aft through the chart room to the hatch. He looked out through the heavy plate window in the door, then came back.

"Helicopter's looking pretty secure." He went to the gyrocompass in the center of the bridge console, where Lincoln was keeping an eye on the heading, and turned the center knob which adjusted the autopilot. "Try that for a while," he said to his quartermaster. "One hundred fifty-five degrees. We'll head a little more east, run with the waves a little more."

Another big swell pitched the ship. We all hung on. "Trouble below Tasmania," Watson said; "no land to stop the swells. They're called Cape rollers. Can get huge. Yeah, *The Perfect Storm*—we were rooting for the storm. Load of crap. Nobody's going out on a yardarm in those kinds of seas with a welding torch. Those guys died because of their own greed. Sitting there systematically wiping out the swordfish. What's the difference between that and going to Africa and plugging elephants—bang bang bang."

Two royal albatross soared up past the starboard windows. They seemed to be looking in. They were huge and white, with black on the upper side of their wings—and their wings were something like ten feet across. Without moving them they dived down past the lunging bow and swept almost straight up like kites on strings. Nothing seemed to propel them but the power of sheer intention. Whereas the *Farley* labored and rolled, they swooped with exuberant ease, trailing the long extended wing tips an inch, or less, above the tossing water, rising out of a protected trough as if gravity were for other worlds than theirs, canting over with the angle of the swell, stiff-winged, to skim it without touching, rising to peel off downwind in a sudden capitulation to sheer speed. They widened the arc to swoop down again into the ship's wake, to land without a single beat of their wings. They floated there on the foaming track, watching us chug away, bobbing as neat and easy on the storm as swans on a millpond. Sometimes they did all this in pairs, as did the shearwaters and fulmars, in perfect, terrifying sync—the terror of sheer, heedless, inhuman beauty. A wandering albatross can cover over 9,000 miles in a single feeding trip and in a lifetime will clock ten times the distance to the moon.

Watson saw them. "Longlines—that's what they were using in *The Perfect Storm*. The leading cause of the rapid diminishment of albatross numbers worldwide. They dive on the bait as it goes out and get pulled down and drown. Right now, seventeen of twenty-four species face extinction."

I went down the outside steps off the bridge wing and made my way aft to the stern chains. Geert took a break and came with me. We could hear the grinder whining on steel, and hammering metal. The wind was freshening and cold out of the NNW, and the big rollers were now coming almost on the stern. I'd lashed an old wooden armchair to a standpipe back there. Geert found the other one. We looked out through the chains to the wake and the birds flying over it. The air was bright, windy, but the sun was hidden in racing overcast. In the Bass Strait and along the Tasman coast we had seen gulls, terns, boobies, but now the shorebirds were all gone. There was a small black bird with a white rump that fluttered and cut like a swallow; it looked more like a land bird, but it wasn't. According to Dennis Marks, a graying deck hand who was a seasonal bird biologist up in Alaska with the U.S. Fish and Wildlife Service, they were pelagic, a kind of petrel. We were also seeing now the big, heavy sooty shearwaters that were almost black; and Cape pigeons, black with white mottling, that weren't pigeons at all but also a kind of petrel. These birds, big and small, were completely comfortable with storms. They would stay at sea sometimes for years at a stretch, without migrating northward to more temperate waters in the Southern Ocean winter. There is no more exposed, intimidating environment, except perhaps this ocean even farther south where it began to fill with ice. They could slip among the stiffest winds and most raucous waves with peerless ease. In that way they were like the whales.

The next two days passed quickly. Marc worked morning to night on the main deck welding and grinding on some four-inch steel I beams. Shy Willie was often out of the engine room and in the bosun's locker, working with a will on pieces of steel. The two seemed to be working together. The grinder's gritty sear whined late into the nights, which where getting shorter. The smell of it, the acid smoke of steel, filled the companionways. The sun rose at five and set at nine. The weather held, the seas ten or twelve feet, always out of the north-northwest, the racing overcast unremitting. More albatross accompanied us now, black-browed, wandering, royal. Dennis told me they had special bones in their wings that locked them stiff and took the stress off their muscles so they could glide for hours, days. They never seemed to tire of following the ship. Why did they do it? It wasn't opportunistic: I never once saw them catch a fish in the wake. I liked to think it was because all sentient beings have a measure of curiosity and enjoy company.

At dinner on December 15 Alex announced that we would be in the ice in four days. Joel and Justin asked Ron and me if we would judge a haiku contest to pass the time.

All entries were due Saturday at dinner, in two days.

After dinner we played poker. The J's, Trevor, Darren, the captain. We sat at the biggest booth table and Watson popped the latches on the ship's poker kit, a little briefcase with chips, cards, markers.

It was agreed: five bucks, any currency—Aussie, Kiwi, U.S. No pesos, no euros. No yen.

It was Darren's deal. He had a schooled British accent. "Not working very hard now on the bridge," he said, laying out each card with a peculiar emphasis. "But in four days, then I'll be working hard, sir."

"Working hard?" I said.

"Steering through icebergs at night, that's hard work, sir."

When I lost all my chips I wandered out onto the stern. Electrician Dave and Wessel were back there smoking. It was 2200, ten p.m., and the sky had cleared enough so that we could see a full blood moon rising. The sun had gone over the horizon and was flaying the gray overcast to the west. Cold wind and cobalt water. We were back on an unwavering 176-degree course, and the black ship seemed to be driving head down into the south—plunging, pitching, rolling—with a purpose now almost of her own. It was as if the will of the captain ran through the *Farley* like a pulse.

Wessel, in shorts and flip-flops in the raw cold, hand-rolled a ciga-rette. "We're in the middle of fuck-all nowhere. We've been lucky with the weather."

Dave pried a lighter out of his checked flannel overshirt and lit up. "Watch the birds, mate; they'll tell you when a storm's coming."

"How?"

"They'll piss off."

The two smoked. Night passed over the reefs of cloud to the west like the shadow of a wing. The clouds lost their color and deepened to match the sea. The moon brightened.

Dave said, "Yeah, you can tell everything from the birds, eh?"

The next morning I got a cup of coffee from the big pot in the green room and wandered out onto the main deck before breakfast. I liked to fix a waypoint on my GPS first thing, so I'd have an independent record: *12/16 0700 54°49'S, 148°08'E.* The day was cold, solid gray, and the wind had picked up and shifted to the northeast. It felt some-how as if we were entering a different marine landscape. I glanced at the steel T beams lying on the deck. Marc had been working on them for the last two days. The steel struts now formed a—blade. A seven-foot triangular cutter, polished and sharpened at one end, exactly like a can opener. The can opener I had read about. It was at that moment that I fully realized this was not a game.

On the bridge, even the quartermasters were working on their poems. Watson sat in his high chair, and between entertaining the watch with excerpts from his book in progress, *The Aquatic Ape*—an elaboration of the theory that modern humans evolved from an amphibious *Homo* ancestor—he scribbled on a sheet covered with haiku. Geert, drawing, as always, at the chart table, moved his head rhythmically and murmured to himself.

After lunch, a dense fog rolled in over the water and the wind and temperature dropped. The following albatross glided out of it, white on white, and were absorbed again. If they were ghostly before, the fog gave their flight the barely acknowledged constancy of a fixed idea. The mist seemed to settle the swells like a calming hand. The wash of the bow was amplified, as were the hammerings on the deck.

Watson said, "Some sailors hate fog. I love it. Nice and calm." He planed his hand out ahead. I thought, Of course you do; you are all about evasion and surprise.

He said, "A few years ago the USS *Enterprise*, the aircraft carrier, was coming into Puget Sound in heavy fog. They radioed ahead, 'This is the USS *Enterprise*. You will change your course.' The radio came back: 'Be advised that you will change *your* course.'

"'This is the warship USS *Enterprise*. You will change *your* course.'

"'This is the Puget Sound lighthouse. Please be advised that you will change your course immediately.'"

I asked him if he had ever been mistaken for a warship.

He said, "In 1991 the first mate took the *Sea Shepherd II* from Trinidad up to Virginia. One of the crewmembers' father was in the navy. He said, 'I know a harbor pilot can give you a cheap berth.' The pilot boards the boat, takes it into harbor. The crew noticed it was all navy ships. A truck pulls up, gives us a water hose. Gave all the crew passes—navy passes for three days. I finally show up, get to ship, couple of officers come down, say, 'You guys British navy?' We had a black ship, a British flag. I say, 'No.' They say, 'This is a navy facility. You can go to jail for being here.' I say, 'Well, we've been here three days.' The navy had a problem explaining how we got away with three days unnoticed, so they didn't charge us.

"In 1991 or 1992, the *Edward Abbey*, a former Canadian coast guard PT boat, was in one of Tom Clancy's minor movies. We were taking 'nuclear missiles' off our boat and putting them on a bigger ship. We were white with Russian flags, lots of papier-mâché guns. I mean big guns. The actors were all on deck in Russian uniforms. Two of them seasick in the harbor. We pulled into Marina Del Ray, a coast guard station right there, and everybody stood around; nobody noticed. Nobody even checked us out. I guess in LA you get used to seeing all sorts of weird things."

He laughed. He went over to the chart table and Geert pulled away his colored pencils deftly. Watson moved his plump hand over the small-scale chart that covered all of Antarctica.

"Antarctica is bigger than Australia—like New York to LA going from here to here." He pointed to the coast of Wilkes Land and then to the northern end of the Antarctic Peninsula. "This is 600 to 700 miles of solid ice." He spread his fingers across the Ross Ice Shelf, which lay to the east of our present course.

I said, "You put out a press release in Melbourne saying that you were cooperating with Greenpeace. You told everyone that you were heading south, then west to converge on their ships. But we're going east of south if anything."

"That's what we want the Japanese to think. But I think they'll be hunting over here"—he pointed at the Ross Sea, the George V Coast, off to the east of us—"to avoid Greenpeace."

It was big, all right. More coastline than the Atlantic and Pacific coasts of the lower forty-eight, plus the entire coast of Alaska. I was beginning to think Watson had no idea where the Japanese were.

The fog rolled across the decks all afternoon like smoke. It roiled before the bow, veiling the swells that churned away to the southeast, which seemed to be where we were heading. It erased our wake behind us. I thought it was a good metaphor for Watson's strategy of deception. He was turning the *Farley* into a black ghost ship, sailing her into a fog bank and off the map. In the fog and overcast, nobody would know where we were, not even conventional satellites. How often do you sail to Antarctica with no clue of where you are going? Just southward, latitude by latitude, pulled by the force of a single will. It made me feel uneasy. We were flying under the radar, so far under that we might disappear without a trace. We were all alone. Strike an iceberg, tear a large hull breach, go down in two minutes. I was sure we had an Epurb on board, the emergency radio signal beacon; Aultman said he had one on the chopper. Did it work? On the

Farley, I gave the odds as fifty-fifty. Would Watson activate it? What good would it do in the empty vastness of these seas?

58°51'S, 148°44'E

December 17 broke with cold misting rain and following seas.

A sense of taut excitement, inexplicable, ran through the ship. Many of the haiku entries were about anticipating battle.

> *Wait, what is that sound?*
> *Hundreds of drowning whalers!*
> *Quick, throw some bricks!*
> —GEDDEN

Just after breakfast, Trevor climbed to the monkey deck above the bridge to test the topmost water cannons. A thwomp like a sumo wrestler hitting the mat, and then the windows of the bridge were covered with a deluge of water. Old pipes bursting. A valve shut somewhere, and again only the fine drizzle running down the windows and an empty black sea.

Trevor breezed onto the bridge holding a monkey wrench and said, "We could hook the diesel pump right to the water cannons. Spray the factory ship down, tell them to get off, they launch the lifeboats—then toss a couple of Molotov cocktails. That would do it." He smiled so that I knew he was joking. Maybe. Later today he'd get Marc to weld new standpipes under the guns.

Marc, Willie, and Steve readied the can opener for deployment on the starboard bow. Marc, cheerful in his bright slicker in the near-freezing rain, welded a reinforcement plate on the deck just behind the slot in the bulwarks.

The chopper pilot, Aultman, was in the radio room, bent over the desk, which was covered with pieces of a ham radio. Since the marine radio in the chopper didn't work, he'd try this. He was so anxious to fly that he had to keep his hands busy with something. He'd been talking with the captain, and the plan now was to stop the ship and launch the chopper for a long sortie ahead, then move. That way they could save fuel and stay out longer. It seemed Watson was hoping the Japanese might come to him.

At 0900 Jim Pacheco reported seeing a perfectly round lemon slice in the water.

The news swept the ship.

At 1040 Lincoln reported five mic keys on channel 13. That's what you hear if there's a radio transmission too far out of range to make out voices. A chirp or cough. Somebody was within 100 miles.

Allison pulled people into private meetings in the captain's state-room—the captain himself, Alex, Geert. Her inner circle.

Moods pervade a ship from engine room to bridge and can change as fast as the weather in the Southern Ocean. Secrets are hard to keep, and if they are kept, an uneasy awareness of an obscure significance, of a withholding, is also universally felt.

Most of the crew knew that Allison and Watson had something up their sleeves, though not specifically what.

I asked Watson where he got his authority to weld up can openers and ram ships.

"The UN Charter for Nature," he said in his offhand, easygoing style. "Article 21, under the section for Enforcement." He got up, went to the international treaty bookshelf, and came back to the bridge. "Here it is." He'd opened one of the books to a dog-eared page.

"The United Nations World Charter for Nature, ratified by the UN General Assembly in 1982, states under this section on imple-mentation: 'Any nongovernment organization, individual, or nation-state is empowered to uphold international law, specifically and especially in international waters.'"

"Do you think they actually meant you could go out and damage ships?"

"You know, I was arrested off the Grand Banks in 1993. Canada arrested me, but not the Cuban trawler that was fishing illegally. They tried me on two counts of mischievous damage to property, two counts of threatening life. The mischief counts, I faced two times life plus ten. More than the O. J. trial." Watson chuckled, but his dark eyes were dead serious.

"I was the only witness for the defense. The Newfoundlanders hate me for my stance on sealing and other things, but they hate the government more. They were the jury. My lawyer said that we are going to say we did exactly what were charged with. The color of right. I acted in accordance with what I thought was proper and lawful. Our defense was that the UN Charter for Nature gave us the right. We said, 'Did Canada sign this?'"

He folded the charter. "They acquitted me. Here—" he meant Canada, "and in the United States you do have the right of citizen's arrest. If the ship was to cause—" he backed up. "A ship is like a corporation. Under international law the ship can be charged without me being charged. A Japanese flagged vessel, a Canadian flagged vessel, in Australian Antarctic territory—it's complicated. Where does it go to court? But say the Australians charge us. That acknowledges that they have the power to stop the Japanese. But they don't; they stop us."

"Why go to such lengths to stop the whalers? The Japanese say there are 760,000 minke whales. They say it's sustainable."

"That's bull. The IWC says it could be 300,000 and they admit they don't know. All great whales are endangered now. The bowhead whale, the Pacific gray whale, would now be extinct except for conservation laws. The remaining whales are threatened by whaling, heavy metals, global warming, pollution, longlines, drift nets, low-frequency sonar, ship strikes; 300,000 whales and small cetaceans are killed every year by getting tangled in fishing gear or as bycatch. They say the great whales need at least fifty more years to recover, but even then their survival is tenuous. The earth's whales simply cannot endure another period of open commercial whaling." The bridge was full now—the seats along the back, the places by the windows along the side. For a lot of the crew, listening to Watson was like going to church.

"You know what the Japanese delegate Tadahiko Nakamura said to me at the 1997 IWC meeting in Monaco? He said he didn't care if all the whales died. He said his duty was to his family, his company, his country, and that it was his duty to harvest all the whales they could before they were all gone. 'Realize maximum profit from them before they go extinct' are the words he used." Watson took a deep breath. He wanted to shift gears. "Hey, somebody get that kid Watkins. The one who spent his twenty-first birthday in solitary."

A minute later Jeff stood in the doorway, rubbing his cold hands together—he'd been wire-brushing rust on the aft deck. Watson appraised him.

"I hear you're supposed to be an expert on comics. We'll have to have a debate."

"That's one you'll definitely lose, captain."

"What was Spiderman's first girlfriend's name?"

Watkins didn't blink. "Gwen Stacey. What was her father's name?"

"Captain George Stacey. Who else did he date?"

"He had some run-ins with Betty Brandt at the *Daily Bugle*. He was invited to join the X-Men. Also invited to join the Fantastic Four."

They were off. Half an hour later I had them even at four apiece.

Watson said, "I always liked Green Arrow better than Green Lantern."

Jeff agreed. "Yeah, he didn't have any super powers, but he was always bagging and carting off the female superheroes . . ."

I went out the bridge wing and down the outside steps. A wandering albatross glided by in the fog, and I wondered if I were in a weird dream. Watson had talked about the vulnerability of whale populations. There was the question of why the Japanese were down here whaling at all. Even from an economic standpoint it made little sense.

One fin whale might bring close to $200,000 on the wholesale market, and much more at retail; but in 2004 the fleet had brought in $78 million, which the Institute of Cetacean Research (ICR) said covered only 90 percent of its expenses. A large government subsidy, ranging between $8 million and $73 million annually, helps keep the ICR afloat. Masayuki Komatsu, counselor of the Fisheries Agency of Japan, said at the annual meeting of the International Whaling Commission in 2001, "It has been ten years since we started scientific whaling. Though the surveillance study has made progress, the ten years has been a dark period on selling the by-product in good quality for reasonable prices."

The market for whale meat has gotten dimmer since then. Recently I had read two surveys that found the Japanese appetite for the dense red meat to be at an all-time low. A survey conducted by Britain's leading research company, MORI, in 2000, found that only 1 percent of the Japanese public eat whale meat once a month, and that only 11 percent support whaling. The *Sydney Morning Herald* reported a glut of unsold whale meat—a record 4,800 tons stockpiled in freezers at the start of this whaling season. The price had fallen from fifteen to ten dollars per pound. There was such a surplus that it was being used in pet food. The meat was available in a few expensive niche restaurants, and was being shunted by the government into school lunch programs across the country in the hope of encouraging consumption. The Japanese government was also spending $5 million a year on programs to boost sales, according to the *Washington Post*. There was even a Whale Cuisine Preservation Association—all this at a time when numerous environmental organizations were reporting that whale meat could be dangerously high in toxins such as mercury and PCBs.

In the face of opposition, the Institute of Cetacean Research pointed to the fact that whaling was an important part of traditional Japanese culture. Journalist Norimitsu Onishi wrote in the *New York Times*: " . . . to unify public sentiment behind whaling, the government promoted the argument that whaling was part of Japan's cultural heritage and that it was being threatened by the West. . . . The argument resonated in a country where many feel that traditional culture has been lost in Japan's confrontation with and then embrace of America . . ."

But the truth is that only isolated coastal communities such as Ayukawa and Taiji had been hunting whales for centuries. After Commodore Matthew Perry forced open an isolationist Japan's markets with his four gunboats, "the Black Ships," in the 1850s, Japan began to adopt Norwegian and American commercial whaling methods;

Japanese, especially in the western part of the country, began to eat more whale meat. But the mainstream Japanese public didn't become acquainted with whale meat until the food shortages in the period just after World War II, when MacArthur encouraged the Japanese to eat it. And according to the MORI poll, "Virtually nobody fears Japan's cultural identity would suffer greatly were whaling to stop."

Although the first Japanese whaling fleet went to Antarctica in 1934, it wasn't until 1948, when the whalers became part of the Japanese fishing powerhouse Nissui, that the meat was canned and aggressively marketed. Two other large Japanese fishing companies, Maruha and Kyokuyo, also whaled in the Southern Ocean through the 1960s; and in 1976 they merged to form one whaling company, Nippon Kydo Hogei. It is estimated that the three companies killed nearly 500,000 whales between 1929 and 1986.

In 1987, with the implementation of the IWC's moratorium on commercial whaling, Japan established the nonprofit Institute of Cetacean Research, under the supervision of the Fisheries Agency, to conduct its "research" in the Southern Ocean. Japan's Research Program in the Antarctic (JARPA) was born. To most observers this was a program designed to circumvent the moratorium and keep the commercial whaling fleet afloat until the moratorium could be overturned. In the eighteen years of JARPA, the scientific committee of the IWC had lodged no fewer than twenty objections to Japan's "lethal research." (At the meeting of the IWC Scientific Committee in 2001, thirty-two scientists from around the world submitted a paper claiming that JARPA lacked scientific rigor and would not meet the minimum standards of peer review.) Also in 1987, Nippon Kyodo Hogei became Kyodo Senpaku. The three big fishing companies—Nissui, Maruha, and Kyokuyo—each owned a one-third share. Kyodo Senpaku now had a lock on Japanese whaling. The ICR contracted exclusively with this corporate giant to conduct research whaling in Antarctica and the North Pacific. The ICR set the price of the meat, or "by-product," and consigned the company to sell a portion of the catch to licensed wholesalers for a commission. Nissui and Kyokuyo continued to package and sell the meat. The meat that cannot be sold on the open market continues to be stockpiled, and sold by the government at discounted prices to schools, hospitals, and other public institutions, and to local governments to promote the eating of whale meat.

The ICR has never brought in enough revenue with the sale of whale meat to cover its expenses.

So why, if whaling was essentially unprofitable, was the Japanese fleet in Antarctica attempting to take more whales than in any other year since the establishment of the moratorium—and in the face of so much international outcry and pressure?

Some international observers believe it is exactly this pressure that is responsible for Japan's recalcitrance. As quoted by the BBC, Hideki Moro, a top official at the Fisheries Agency, recently said, "If the current ban on hunting whales is allowed to become permanent, activists may direct their efforts to restricting other types of fishing."

Japan is the world's largest consumer of seafood. According to a report by the Carnegie Endowment, 40 percent of the protein in the Japanese diet comes from marine species. And with these sources threatened by shrinking fish stocks and collapsing fisheries worldwide, the Japanese may be especially sensitive to criticism of any of their harvests. This may be why they are so anxious to blame whales for the decline of fisheries.

"As long as officials present the issue [of whaling] as one of Japan being bullied by the rest of the world," the BBC quotes Greenpeace's John Frizell as saying, "they can probably keep the Japanese public behind them."

All the indicators seemed to suggest that the Japanese people don't particularly like whale meat, that they don't think it's an important cultural value, and that the industry is facing a growing surplus and a tougher time making money. If, in the end, the Japanese whaling fleet was now in the Southern Ocean Whale Sanctuary gunning for whales because of notions of national pride, it seemed to me that members of vulnerable species shouldn't be hunted. Not for "research," when nonlethal means were more effective; and not, especially, on a commercial scale.

No darkness now at night. The sun set just after midnight but left an uncertain twilight that lingered until dawn four hours later. Near midnight I took my every-three-day, three-minute shower in the cold stall, then went to the green room to watch a movie. Chris Price was regaling the crew with a story about a trucker who pulled into a weigh station with a funny stench coming from his trailer, and how they found a dead whore back there. "It's true, I swear!" he declared. "You could look it up." The engineers and the Earthfirsters wanted to watch *Death Race 2000*, and the J. Crew, the French, and the Canadians voted for *Jughead*. Geert stepped forward, gangly, bushy beard, in his biker vest and heavy black work boots. He'd spent all day drawing sweet pictures of big-eyed seals.

"How about *Miss Congeniality*?" he said shyly, holding out a DVD.

A shocked silence settled over the cabin. Everybody blinked at Geert.

"*Miss Congeniality*? Dude!" Justin blurted.

Marc spoke up for his countryman. "Geert has got a fat love crush on Sandra Bullock. Don't bust his balls."

Geert winced. Somebody grabbed the DVD and slid it in. Everybody was fond of the ship's artist.

10

Ghost

December 18, 2005

62°26'S, 149°24'E

The fog descended again on December 18 and the ship moved within it like a scent hound. In this case her nose was two radar which spun silently on their masts above the bridge. The water glassed off black, a shifting patch of road ahead, and the swells pushed the *Farley* gently from behind. At breakfast we were at 62°S, still running down the 149th meridian toward the Virik Bank, and we were now only about 250 miles off the coast of Antarctica. We were entering the latitudes of ice. Visibility was less than 300 yards. The quartermasters watched the pulsing green screens with avid attention. Not only were we entering Antarctica; we had also passed the northern edge of the Japanese hunting grounds at 60°S.

During dinner we encountered our first iceberg. It loomed out of the fog, at first ghostly, condensed from the same whiteness as the air; then, closer, a solid alabaster wall crazed with blue crevices. It was a cliff 100 feet high, a quarter-mile long, flat-topped like a floating mesa, and the swells washed against its base with a sound like low laughter. They carved a hollow undercut of a fierce and vivid blue. The color was almost the blue of shallow reefs seen from the air, but colder, harder, unforgiving.

After dinner Ron and I announced the winners of the haiku contest. The entire crew gathered, except for Alex, who was on the bridge. We read aloud the honorable mentions. Then Emily opened the envelope and announced, "The third-place winner is . . . Geert and Marc!"

Water water ice
nice and cozy sleeping on the fucking bridge
giant iceberg, Boom!

Applause.
"The second-place winner is . . . Envelope, please . . . Trevor!"
Ron recited: "Marine Engineer's Haiku":

Precious. Ferocious.
The perfect ramming machine.
For the sea! Freedom.

Then the winner:

Hobart. Beer falls in.
Splash. That was rum. Boat lost too.
Saving whales is hard.

The winning poem was anonymous, but after some uproar Kristian, the videographer from Vancouver, stepped forward to receive his hand-painted plaque.

At 2200 I was in the middle of an intense head-to-head match of Texas hold 'em with Jeff, when Mia tugged my sleeve and summoned me to the bridge. It was crowded. All the officers were there. I glanced at the GPS giving our position: 64°19.95'S, 149°49.69'E.

Alex stood over the main radar screen. Watson was in his chair. The fog closed around the bridge like a shroud. For the first time I saw chunks of ice floating and heard the grate as the bow sloughed one off.

There were three blips on the radar, two moving. The size of small peas. The closest was 6.4 miles off. It was ahead, off to our starboard, about three o'clock, moving almost due north at eleven knots. I noticed that we had swung around to the north as well and were in pursuit. Another blip was at ten o'clock, moving more slowly. The third was behind us at five o'clock, motionless. On channel 13, the chirps of keying mics were constant. Two or more ships were talking to each other.

Had Watson sailed straight into the middle of the fleet? It was too good.

Alex had slid a cursor over two of the blips and marked them with tracking circles. Around the third—the biggest, the one moving fast— he had marked a box. The marks gave him the speed, bearing, and range, or distance, of the blips inside them. From the box extended a line. He had the range set so that it projected the ship's path half an hour out.

"I'd say they're not icebergs. I'd say they're ships," he said to Watson.

"I'd rather go for the one that's moving."

Alex picked up the hydraulic rudder thumb control, and the *Farley* began to swing away from the groundswell to starboard. She rolled as the waves came abeam. The GPS over the console showed the heading at fifty-five degrees. On the radar, the projection line from the center of the circle, the bull's-eye that was the *Farley*, swung to cross the course line of the other ship.

"Intercept half hour," Alex said; "5.2 miles." He handed the helm control to Hammarstedt. "Keep it there. Fifty-five."

"Fifty-five."

"Alert the crew," Watson said. He was very calm. "Get someone in the crow's nest."

Casson said, "I'll go. Should I bring a radio?"

"No, don't bring a radio."

"Port five," Alex said.

"Port five," Hammarstedt repeated.

"They're moving faster. Port five."

"Port five."

Alex was swinging by degrees to port so they didn't beat us across the bow.

"Hey, Alex," Kalifi said, "you want me to get the Zodiacs ready?"

"Not till we find out what it is." Watson glanced up at the secondary radar over his head. It was set so that moving objects made a smear, so the operator could quickly tell the direction of travel. "I see he's moving this way."

"They haven't changed course at all. Three hundred fifty-three degrees. We are heading twenty-five."

The radio stuttered with static. More mic keys. The neighborhood was certainly getting crowded.

"Is it channel 13 they're coming on?" Hammarstedt said.

"Yes," Watson said. To Alex: "How far is it?"

"Four-point-nine miles."

"That's the only one that's moving, right?"

"Yes."

Silence. The dense fog in no hurry as it rolled across the bow. Light creaking of the bridge timberwork. Something swinging rhythmically against a bulkhead in the chart room. 2213 by the GPS clock. Less than fifteen minutes had passed since I arrived on the bridge. Time was at once compressed and slowed. A lot had happened in a very few minutes, and yet it was all unfolding with the choreographed slowness of a Noh play. I was not used to the pace of the sea.

"Port twelve."

"Port twelve."

I leaned over Alex's shoulder. "We're trying to intercept this one"—he pointed to the one in the box, the only one presently moving and now at about two o'clock— "and pass this one." He indicated the blip in the circle at about eleven o'clock. "Get two for the price of one. We're heading almost toward it. It's about four miles away."

Visibility was less than 300 yards so I watched the radar. Alex pointed to another blip that had just appeared, also about two o'clock, but farther out. Possibly four ships together.

Watson said, "Thirteen again."

"There's been static on that channel for the past four days," said Hammarstedt.

Alex: "That's a ship. No doubt." Skinny Alex had every sinewy ounce focused on the screen.

"Probably a Greenpeace boat trying to figure out if we're a whaling boat, trying to sneak up on us," Watson said. "Well, in two and a half miles we'll know if it's an iceberg or a factory ship."

Watson had been preparing four years for this moment, and he seemed almost too relaxed.

Alex said, "Could be three icebergs and a boat, of course. Fifteen minutes."

Allison handed Watson his Mustang suit, which he stepped into. It was orange and black, and it said "U.S. Coast Guard" on the back. "Got it on Ebay," he said. He shrugged into the arms. "I don't believe it's 10:30." Outside was the same grave murky light we'd had all day. He got back into his chair and glanced at the radar above him.

"Is the other guy changing? How close can we get?"

"About a mile. He's moving too fast to intercept."

"If you're absolutely sure that's a ship, maybe we should get a Zodiac in the water."

Alex sprang. "Yup," he said simply. He straightened and twisted away from the screen. "OK," he barked. "Get a Zodiac in right now!" His hand came down hard on the wood of the console. "Go! Now!"

Kalifi shoved open the door to the outside steps. Colin was behind her. Watson said, "Peter, who do you want on the boat?"

Alex cut in. "Colin." The bosun's mate turned at the hatch. "We'll stop the boat when you get in. Make sure you take a prop fouler."

"Got it."

"Have you got a carabiner for the radio?" Hammarstedt said.

Alex snapped, "He's got a biner, don't worry about it. Go!"

Half the bridge emptied. Through the windows we could see crew pouring out onto the deck, whipping the tarp off the port Zodiac, slinging the harness. Kalifi clambered to the davit controls and the crane swung over. Mist rolled over them in veils.

"Tell them to bring the radio for emergency," Watson said. "Don't use it except in an emergency. In 1976 we lost three Greenpeace Zodiacs in a fog like this. Of course we didn't have radar. How are we going to direct them where to go?"

"In this soup it's fucking hard to find anything—they should be wearing Mustang suits! Lincoln, go down; tell them to put on Mustang suits. Not to launch it till we give them the sign. We'll slow down." Alex turned to Watson, "The ship is pretty much on a northbound course. They've got a GPS and a radio."

"Have the radios off unless there's an emergency?" Hammarstedt said.

"No, the radios on. Just don't transmit."

"The problem we had in Hobart was they were transmitting."

Watson said, "That's because Inde had the damn thing on transmit all the time—that thing in front is definitely an iceberg. Another flat-top, it looks like." It was. It loomed out of the fog, thirty feet high, maybe 400 long, the same dimensions as a factory ship. But the other blip was still moving at eleven knots on a course of 353 degrees, unwavering. Icebergs didn't do that. Of course, whoever it was had radar at least as sophisticated as ours and could see that at our present speed we would not force a collision.

Alex had shoved down the sliding window in front of him and was watching the deck like a fierce hawk. "They got the Mustang suits? OK." Lincoln, Colin, and Gedden were in their orange suits, ready to go. Colin tossed a small buoy attached to a coil of steel cable into the Zodiac—a tangle line for fouling a ship's prop.

Alex, out the window: "Get someone to pump that pontoon up!" He pulled his head back in. "There's like six people standing there. They can pump it up right now." I hadn't noticed, but the right pontoon of the Zodiac was wrinkled and sagged like a flat tire. I looked out the window of the door onto the bridge wing and saw the thermometer marking half a degree above freezing. The colder air contracted and it would make the whole Zodiac slack.

Alex yelled, "Get three or four people in there!"

Watson said, "Better take a second radio as a backup." He was still thinking about the safety of his crew. It would be easy to motor off in this fog, and never be seen again. Beyond about a mile, a Zodiac, even on the primary radar, broke up its signature and was lost in a swirling constellation of wave tops and chunks of ice. The chopper could not take off in this soup. I realized how vulnerable any of our craft would be—Zodiacs, and FIB, and chopper alike—as soon as they got out of sight of our own little mother ship.

Alex said, "We can follow them." He craned around to Mathieu, the French filmmaker. "If you're on time you can get in—hey, get them a handheld compass—no, it's pointless. Magnetic south." Hadn't thought of that. We were within 350 miles of magnetic south and all the magnetic compasses were haywire. The south magnetic pole was about 1,500 miles north of the south pole and shifted year to year. It did not have nearly the force of magnetic north, but it wreaked havoc on compasses that were nearby. The ship's binnacle, a big dial compass, was about sixty degrees off at the moment.

Watson yelled out the window, "Somebody take care of that radio! It's on the deck of the Zodiac!"

The outside hatch swung open, letting in a gust of cold. "Jeff's ready," Lincoln said. "They're ready to go."

Watson leaned out the window and yelled, "Anybody loses a radio, he'll never get on a boat again!"

Alex: "Get the ladder ready!"

Watson: "Colin, keep an eye out. We're going to follow you."

We passed the iceberg close enough to port to hear the wash of the waves against it through the open windows. The blue at its base was the only color in a movie that had gone black-and-white. This was a sea story that might have taken place in the early twentieth century, or in the nineteenth. No battles unfolded this slowly anymore. No targets were this uncertain, no boats could be lost so easily. No disappointment in a handheld compass was so stinging. *The radio the radio the radio.* Watson kept repeating it. But what good would the radio do if they got out of range? Or if they could not work the GPS and tell the ship where they were?

Almost on cue, Gedden burst onto the bridge. "The batteries on this GPS are shot."

That figured. Nobody knew exactly where there were extras. I tore down the three flights to my cabin and retrieved my own GPS. I handed it to Gedden. "This is the compass screen. Here's your current position. It's waterproof, so don't worry about that. Here's a biner, maybe you could clip it somewhere."

He tumbled down the steps.

Watson said, "How far is this guy now?"

"About two and a half. When we get to one and a half, we'll launch. Tell them ten minutes."

Hammarstedt called down, "Zodiac, we're launching in ten minutes!"

Just then Emily rushed in, flushed, her ponytail flying. "They need a flare. Where's a flare?" She was hyperventilating.

"Over there."

"Two flares, yeah?"

Alex shrugged.

Watson said, "There's a French base directly to the south of us. These guys are heading due north. I bet it's a French supply ship."

Alex turned to Lincoln, "Tell them to wait up a little bit. I wanna get a little closer. Tell them to follow a course of due North. Zero-zero-zero.

Watson said, "Mia, hail the ship in French on channel 16. Say, 'French supply vessel.' Repeat it. Don't identify us."

Mia took down the mic and spoke the words, repeating the phrase half a dozen times.

Silence. The phantom had no voice. Watson waited. We could hear the ice chunks thudding off the bow. "Nope. No answer. If they were a French ship they would have answered us."

Alex peered ahead through the dense fog. "It's pretty risky, Paul," he said.

They couldn't even get a load of beer from the Hobart dock without nearly sinking. We weren't in any harbor now, we were in the remotest waters on earth. Whoever was ahead of us did not want to communicate. In a vast, lonely ocean where it was a courtesy to hail a fellow ship, they were not answering our call. They were making tracks away from us and they were not slowing down to see what or who we were.

"We can follow them," Watson said, meaning the Zodiac.

"OK, they can launch at will. Get it under four knots."

Watson said, "Tell them it's at one o'clock right now, two miles . . . Just get going straight ahead! Slightly on starboard side! Let's go!" He pulled back the chrome lever that adjusted the propeller pitch and the *Farley* slowed.

Alex called out, "Launch at will."

Kalifi, on the winch, shouted "Good luck!" and raised the hook. She was wearing her pajama bottoms in the freezing mist. The harness went taut, and the Zodiac lifted out of its cradle. Colin, Jeff, Gedden, and Mathieu were in the boat. Kalifi swung it out over the rail and began to lower. Just then, young Luke and Willie Tatters rushed across the deck and piled into the Zodiac like a couple of puppies. Kalifi lowered the boat to the water. For a moment they were being dragged, bouncing against the slow-moving hull and then they unhooked, started the outboard, and gunned off into the fog. Except that they were gunning very slowly, as if the fog itself were resisting them. Even I could see they were overloaded. As slowly as they went, in less than a minute they were swallowed by the fog.

Watson swore. He revved the *Farley* back to full throttle. "Who's on that boat?"

Lincoln said, "I told them three, max."

"They're going the wrong direction!" Alex said. "OK, they adjusted; if they keep on going straight they should be all right."

Watson fumed. "We're going to have drills on this!"

Alex was watching the radar intently. "I've got them barely," he said. "Not very good."

Watson said, "Doesn't anybody have a space blanket? Best thing is they can bring that with them and unfold it to reflect the radar." It was too late for that. They were dangerously heavy and they were about to disappear from their only safety.

"I've lost them," Alex said. "I have absolutely no fix on them." He straightened and looked at Watson. "Before they intercept they're gonna be four or five miles out, and then . . ." What he meant was that they would be lost to us. Four or five miles in this ocean, blind, could be a few miles over into the Next World. Everybody looked at the captain. The bow struck ice and it scraped along the hull. The low swells rolled onto the port bow and there was the shirring rush as the *Farley* rode over them. Kalifi came through an outside hatch and pulled it shut.

"They should be almost on them," Watson said. "It shouldn't take long to do two miles." He reached up and slid the mic out of its bracket anyway. "Zodiac, do you copy?" No answer. He looked at Kalifi. "Did they turn them on?"

"I think they understood not to turn them on unless they saw something."

"Zodiac, do you copy?"

Nothing.

"Obviously, they don't have them on."

We all watched the captain. There was ice in the water. The seas were calm now, but in these latitudes the weather could change fast. In the time it took to strip a line of hanging laundry, waves could be washing over the deck. It wouldn't take much to swamp the little rubber boat, especially as it was overloaded. In minutes, in their unsealed Mustang suits, the Zodiac crew would be dead. The helicopter would be useless; it was even more dangerous in low visibility.

"There!" Straight ahead, through a rent in the fog, was the stern of a great ship. It seemed terribly far away. The shape of a stern, and the suggestion of a towering superstructure like an empty gate, whiter than the fog around it. Then it was gone.

"That is a whale ship," Watson said. "I think it's the Japanese 'researcher.'"

No sign of the Zodiac.

Watson turned to Hammarstedt. "Want to tell them to start the air on the air horn? That's how we got back to the Greenpeace boat when I was in the Zodiac. Well, these guys are right in front of us now, two miles."

"This ship has been constant," Alex said.

The captain keyed the mike and tried again to raise the Zodiac. Nothing.

"Call them on 16," Alex said. "See if they answer on 16."

Watson did. "Zodiac, do you copy?"

"Copy." It was Gedden's voice, out of the static. Relief.

"Return."

"I'm sorry, what?"

"Is it clearing up where you are?" Watson said. Ahead, the ship was visible again for a few seconds, more as an absence of miasma, a kind of concentration of gravity in the vague shape of a stern.

"It's clearing up."

"That is a ship; there is a ship ahead of us. It's a whale ship."

"Do you want us to proceed in the direction your bow is facing?" Gedden asked.

"Roger."

That seemed like a strange question. But a minute later the Zodiac appeared. It was coming up behind us. Overtaking us to port. They had done a complete circle in the fog. No wonder they hadn't caught the whaler. I didn't get it. They had a GPS with a satellite-guided compass.

Geert said, "Do you think they can see us?" The whole bridge erupted in laughter. I guess everybody was relieved.

"Passing on port side right now," Gedden said without irony. "We can see the ship. I'm heading 030."

"Yeah, we can see that," Watson said drily. "Next time you take a coffee break let us know. Go ahead."

It was 2323, almost midnight. The pursuit had lasted almost an hour and a half. The target ship was dead ahead of us and gaining. At 2.3 miles we couldn't see it anymore. Gedden was motoring off thirty degrees to starboard, heading for empty ocean. The captain got on the radio, frustrated, and reiterated the course: dead ahead, 000 degrees.

"We're not moving very fast," Gedden answered.

"It would help if you didn't have six people. Return." That was the final order.

Allison said, "You don't think it's the mother ship?"

Watson said, "No, it's a research ship, white."

"It's going north, too."

"Yeah, it's coming from the French research station. There was a supply ship scheduled to leave this morning. Going back to Hobart." Watson exhaled. "We have to have a dedicated crew for each of these boats."

"We do," she said. "Unfortunately they're not in it."

I thought the ghost ship could have been the *Nisshin Maru*. There was no other reason for its eerie silence. We were two miles away from it. We were clearly trying to close the distance. Its officers could see that as well as we could, on their radar. We had hailed them in French. Probably there was no one else in these seas for thousands of square miles. They maintained their silence and sailed off. Watson first suggested it was a research ship, then a supply ship. It did not look white to me. The superstructure looked white and the rest looked like fog. I thought our captain was in denial. I could understand it. To go a whole campaign in 2002 in these same waters and find nothing. To prepare for four years for this attempt. To come almost straight south from Hobart and run smack into the factory ship in the middle of a whiteout. That would have been a miracle. And then not to be able to do anything about it. Because the *Farley* was too slow. Because the outboards on the Zodiacs were underpowered and worn out. Because the crew was incompetent. It was too awful to swallow. Watson had to believe that big ship was somebody else.

What would he be able to do the next time? He had mentioned samurai charging guns with swords. In World War II the Polish cavalry charged elements of the German Fourth Panzer Division at the battle of Mokra, and the Germans withdrew. But what if they had charged and the tanks had simply gotten in gear and moved off down the road, ignoring the peppering of pistol shots and the clang of lances?

There was one thing. If they could get the Zodiacs sorted out they should be able to run up on the stern and throw out a prop fouler. That is, if we ever got within a few miles of the factory ship again.

When the Zodiac was at last swung back into its cradle just before midnight, Luke and Willie climbed out of it with less enthusiasm than they'd climbed in. Gedden came onto the bridge, sheepish, still in his Mustang suit, his thick glasses steaming.

"What were six people doing on that boat?" Watson demanded.

"I have no idea. I was fucking with the GPS and the radio. I had to get the radio initialized."

"You were right behind us and the boat. What happened?"

"I don't know. My glasses fogged up; I couldn't see anything. I was telling them where to go. I think Colin's understanding was to go out a mile and if we didn't see anything, come back."

They had been terribly exposed. That made sense.

Mathieu said in his heavy French accent, sheepishly, "The first one was going great and then we saw the *Farley* from the stern."

"Well, that does it, I guess. The fog's back." Watson got up. It was the end of his watch, anyway. "Turn her around," he said to Hammarstedt. "Go 168." Twelve degrees east of due south. He was still heading for the Virik Bank.

11

Ice

I was getting into the habit of waking and climbing to the bridge once during the graveyard shift, Alex's watch, between 0400 and 0800. The ship then was mostly asleep. The odd engineer, or a sleepless videographer, or Hammarstedt or Emily, off the previous watch, might be in the green room watching a movie or sipping tea. Or Trevor, who was ubiquitous—tweaking the engine anytime at night, taking a smoke break at the foot of the bridge steps, or at the stern chains, studying the weather or an albatross. Otherwise, the companionways or halls, the decks, and the chart room were empty. The diesel thrummed and growled. The wind and the waves swashed, keened, percussed; the decks plunged and rolled. I explored the *Farley* then, or studied the charts, or stood on the bridge with Gunter and Alex and Gedden, who were by and large a quiet and serious bunch. Perambulations followed regular patterns: forward and aft, mess to stern chains, stern to bridge, bridge to cabin. Then it was usually time for a meal and back to the mess. There was nowhere to go, really, but where the ship was taking us, and that was a mystery, beyond our control. Right now, that was inexorably south, but it would have to end very soon: there was not much more south left on the entire planet that wasn't solid ice.

Early on the morning of December 19, I woke just after five and climbed to the bridge. The ship was plowing due south in the fog, only thirty feet off a towering wall of ice. It was over 100 feet high, clean-edged and flat at the top, pure alabaster white but luminous, as if lit from inside. The groundswell washed against the electric-blue undercut of its base with an explosion of spray and a hollow boom that repeated with the rote warning of a bell buoy. High up against the smooth face, storm petrels flew and veered like so many cliff swallows.

This was not a cliff, and they were not swallows. It was a great, glacial piece of ice, bigger than Manhattan. The wall ran as far as we could see ahead, into fog, and as far behind.

"We've been passing it for over an hour," Alex said. "Over ten miles. The radar shows it out ahead at least as far."

Hard to imagine the catastrophe when this mass of ice broke from its glacier. I went back to sleep. What woke me again, at seven, was a loud thud. I lay for a minute and listened. The porthole was bolted shut, so the cabin was dark except for a trace of fluorescent light bleeding around the red curtain over the doorway. The vibrations of the diesel, usually pitched at full bore, were subdued. The slosh and gurgle of water along the hull inches from my head had changed from boisterous, even violent, to leisurely, if not tentative. Another thud and a long gritted scrape. I dressed and clambered up the stairs to the main companionway. I shoved open the hatch and stepped onto the main deck.

The *Farley* was moving slowly in a dream of fog and ice. All around the ship, hemming it in, were rafts of broken bergs. Some were the size of dinghies, others like barges. Most were flat to the water, pack ice platforms of old snow; others gestured out of the mist like abstract sculptures of animals. Some cupped hollows that filled and refilled with azure as they rode the swell. Some were smooth, old, and stained; others were broken into sharp-edged geometric forms. The *Farley* moved gingerly through them, hewing to a narrow, jagged trail of open water that disappeared into the fog. She shouldered the smaller bergs aside. The water, smoothed and protected, was a deep lake blue. All of it, this whole floating world, rose and fell on the easy groundswell.

White, gray, black, and the singular, remarkable blues. These were the only colors. The birds were the same: pure white snow petrels and little antarctic petrels, ternlike, glided around the ship. The air coming onto the bow smelled cold and sere, and pure—not a trace of smoke, of earth, tree, rock, grass.

The place did not seem to belong to earth, and yet there were the birds, veering and circling, to remind you that you had not untethered from the planet. In the dark water, seven or eight penguins porpoised fast along an edge of ice, flying through the dark water as their fore-bears once flew through air. Dennis had told me that penguins and albatross were closely related. Some of their common ancestors had made their homes on an island free of predators and no longer needed to fly. Both families had a salt gland that allowed them to drink seawa-ter and secrete the salt.

"Check it out!" It was Casson, our wonkish cook. He pointed to three little penguins who bellied up onto a flat raft of ice and stood together gawking at the high black bow approaching them. They were Adélies. They tottered forward, craned upward, spoke loudly to each other, hustled back a bit. They stood in a ragged line and waved their stubby wings. Just then the bridge cut forward power and the *Farley* ground into the barge of ice and stopped.

I think the penguins thought they had done it. They had evidently never seen such an animal. They rushed, waddling, forward. One tipped over in her excitement and bellied along on the hardened snow and righted herself. They stood together before the bow like emis-saries and flapped their flipper arms rapidly and squawked at each other and at the ship.

Chris Price yelled down, "Where are the whalers?" The one who had fallen down flapped her wings at us, opened her bill, then wad-dled off.

Wessel nudged me. "Look at that." He pointed to a huge brown seal lying on its own raft. Its head looked to be as big as a lion's. "That's a leopard seal," he said. "The fiercest predator out here.

It'll shoot upward, break ice, and grab the other seals. A real badass."

We were at 65°19'S, 150°13'E, about 120 miles from the coast of Antarctica. The Antarctic Circle was about seventy miles to the south, and beyond it was the Cook Ice Shelf. Just west of us was the Mertz Glacier with its huge ice tongue; 180 miles to the south and west was the Australian station at Commonwealth Bay, the windiest place on earth. Wind speeds of 250 miles per hour had been recorded there.

On the bridge, the officers surveyed the scene. We were hemmed in on all sides by undulating ice. It looked intimidating as hell to me, but none of them seemed too concerned.

Alex said to Watson, "Between four-thirty and six-thirty we passed a twenty-mile-long iceberg, then all this broken ice."

Watson said, "When the fog clears we can get the helicopter up to tell us the path."

Chris Aultman brightened. "We can do it," he said, and went aft to the heli deck and began slipping the covers off the chopper blades and engine. He moved with a new vigor. This would be the bird's maiden flight off the ship. Pawel would be his first passenger; he was already out there in his special ops dry suit and all-terrain boots.

The fog didn't clear. We all went down to breakfast.

The *Farley* rocked. We could hear the swell washing over the skirts of ice and chortling in the hollows.

Alex said, "This pack ice could be an ice field, or it could go all away to the coast. I don't think we can go any farther south. We are sending up the chopper to take a look—though in this fog, it's less than one mile visibility."

Watson shook his head. "It doesn't pay to work your way around." He meant to push farther into the pack ice, trying to find a route. "You work your way around and then find you have to—" He stopped. "Is the bow thruster up?"

"Yup, it's up all the way," Alex said.

"We don't want to go any farther into it," Watson said. "This is the edge. Next thing you know the wind shifts and you're locked in like Shackleton."

In the first years of the third millennium we faced the same uncertainties as a ship in the 1800s. Alex agreed. He said the wind could change, and the pack ice could close in around the ship. Now, this morning, there was almost no wind. The fog drifted without purpose like smoke on a still day when low pressure layers it along the ground. The ice, the chunks and rafts, kept their spacing, a jigsaw puzzle spread out by a madman unwilling to lock the pieces and let the picture coalesce. The black *Farley* sat in the middle of it, tempting capture. Whatever current there was moved us all together, slowly westward. In the stillness, even the colors seemed to stay quiet. The one gray leopard seal hadn't moved.

As breakfast ended Watson said he had e-mailed the *Esperanza* two days ago with our position. "Can't get an answer from the Greenpeace people," he said. "They said they wanted to cooperate." He meant certain crew members on their ships.

He showed me the e-mail he'd sent to Shane Rattenbury, the head of Greenpeace's Southern Ocean expedition, who spent time aboard both its ships: the fast flagship, the *Esperanza*, and the slow reconditioned sealer, the *Arctic Sunrise*. It read:

Dear Shane,

The *Farley Mowat* is patrolling east of the 130-degree east line. We have reason to believe that the Japanese may be avoiding the area west of the 130-degree line to avoid a confrontation with Greenpeace and Sea Shepherd. We can save you lots of time and money by scouting ahead. What I am trying to find out is how far east are you now and how far east will you be going? There is no need to duplicate the area covered. It is a big area and we should be able to cover it all with three ships.

We will go east and then double back west again. If we spot them we will notify you.

We have posted a $10,000 reward for the Japanese coordinates.

Captain Paul Watson

I wondered again about the friction between the two groups, which were ostensibly working toward the same ends. It had the heat of sibling hatred, the kind of animosity that arises from intimate mutual knowledge and betrayal. Hammarstedt, the Swede, who had been a member of Greenpeace between 1999 and 2002, said he had gone on a Greenpeace ship and that the Greenpeacers sat around talking about Sea Shepherd the way the *Farley*'s crew talked about Greenpeace.

Watson often talked about founding Greenpeace in 1972. He said he left in 1977 to found Sea Shepherd because he wanted to physically intervene to enforce international laws. The story popular with many Greenpeacers is that he was ejected for grabbing a sealer's club and throwing it into the water. The way Watson tells it, "It wasn't anything about crossing a threshold of nonviolence. Greenpeace works with Earthfirst! and fully supported my ramming and sinking of the *Sierra*. I was voted off the board for opposing Patrick Moore as the new president of the board."

At 0950, Trevor started the engines again. At 1028 we gave up on launching the chopper and Watson pushed the *Farley* cautiously into a wide arc through the leads of open water and turned around. We headed back the way we'd come. The fog was still heavy and spumed across the bow.

Watson tried to put the *Farley* east of north at eighteen degrees—why was he so intent on the Virik Bank?—but we quickly ran into more thick ice. He narrowed his course to ten degrees and then eight degrees, forced to port by the thick pack.

"Looks like it's opening up now," he said to Alex. "You never know."

"Not really. The radar is still a minefield." He was right; growlers thumped off the bow.

And then we were up against the great cliff, the sheared continent of ice, and he ran just off it, now on our starboard, and followed the cold wall due north. There were the petrels again flying along the cliff face as if they were country birds hunting bugs, but there had not been a bug in our world for 1,200 miles. Small ships of ice bobbed by on the swell, and one—eerily smooth, old, and stained—which could not find its base to balance, rolled by; continually capsizing and capsizing again.

Before lunch, Kalifi called an all-hands briefing in the mess to discuss yesterday's action.

Alex stood in front of the food counter. With his buzzed head and wasted cheeks, his intense serious gray eyes, and his scrawny neck sticking out of the black crewneck sweatshirt, he looked like a POW. He had told me, "That's one reason I hate the Japanese whalers—for making me come down here again where I'm always cold." He wasted no words now.

"Couple of things went wrong yesterday. Especially with the Zodiacs. I wanna do it going four to five knots in rough conditions. Next time we're gonna do it quicker. We've got to get really good at this." He looked around the room. Kristy, the tiny nurse, plucked at her rainbow arm warmers. Big Marc crossed his arms over his coveralls and pursed his lips. The J. Crew, crammed into the back booth, watched impassively, like kids in the back of a classroom sizing up a new teacher—or like the poker players they are. Chris Price leaned forward as though he wanted to raise his hand and tell a story.

Alex continued, "Too many people in the Zodiac. When the captain says four, he means four, not six. Need somebody to listen to the radio. Communication is essential. Pretty risky—you go out into the fog and we don't see you anymore.

"I wanna go ahead with the teams we talked about. I don't wanna see anybody else on deck. The three drivers all have their own teams. Kalifi's launching teams are posted at the bosun's locker."

The electronic alarm buzzed loudly in the galley. Watson was on the bridge and summoning somebody. Alex exited.

Wessel said, "Each team must have a boat navigator talking to the skipper."

Allison nodded. "When the Zodiacs go out, whoever is captain of the team should take a GPS." She too was spare, hollow-cheeked, wired hot like Alex. Her baby-blue eyes shone with a warrior's zeal. She and Alex made a good pair of operatives. One could imagine that in a place like Taiji, where they'd freed the dolphins, it would have been next to impossible to stop them.

"And we must know how to use it," Wessel said.

Alex slipped back into the mess. He must have run up and down the steps.

Chris Price finally spoke. It was like a pressure release valve. "The navigator can take a GPS with a waypoint to go to—we can program it so it's all ready to go."

"Sometimes you don't have time to do this," Alex said.

Allison said, "We have to make some time because going out in the fog like this you can get lost so fast."

"OK," Alex said. "We'll do a little training, then. There's an instruction book on the GPS. Don't go up there and start touching buttons." He looked around the mess.

"At first we thought it was five ships," Alex said. "We thought it was the fleet. Turns out it was a research ship." That was a stretch. I didn't think anybody knew what the hell it was. "I realize we have a very inexperienced crew. You can't expect somebody to do everything right away. The navigator didn't do that well but we've never followed a ship into the mist. Even with experienced people, we've never done that before."

"Keep in mind where we are," Wessel said. "You're not off the coast of Australia. This is the shits."

"Be prepared to be in them a long time. Could be a long night."

Kalifi said, "The deck doesn't have to be completely cleared. I just want to see people with these three launching teams. I think everybody did a good job as far as launching was concerned."

Alex said, "Absolutely don't improvise. Do it the way we planned."

Gray-haired Dennis, the bird biologist, said, "I saw some cotton get into the boat. That's bad. Cotton kills."

Cotton. Good point. It's a basic tenet of outdoor survival that one should never wear cotton in any situation that is likely to get cold and wet. The stuff clings to the skin and stays clammy, and saps the body heat. Wool is warm when wet, as are all the synthetic undergarments and fleeces. Dennis's comment underscored how green so many of the crew were. Eager and willing and green. Also, how little basic training they were getting.

Alex said, "Looks like we will be launching the helicopter today. But it's got to clear up. If not, we're probably going to have to keep on going till we clear up."

Joel raised his hand in the back. "I have a question about the mission itself. It seems we're unprepared. We're in the dark as to what we will actually do when we meet them."

"We're talking about that," Alex said.

"It seems," Joel continued, "like we need to make preparations so when we do find them—so we're ready."

Steve said, "It doesn't seem the toys are being worked on."

"Basically a steel cable and a couple of floats is something you can put together," Alex said. What were they talking about?

"Aren't we going to see if we have any stuff left over from the last campaign?" Allison asked, turning to Wessel. "Get barrels, get the cable. When we walked away from the meeting last time I thought you would work on it."

That must have been another meeting many of us weren't privy to. Again there was a sense that a lot more was going on beneath the surface than met the eye.

"Nederlander," she continued, meaning the Dutch welder, "is working on his own project."

"I'd actually like your help," Wessel said to Marc.

"No problem."

"Paul has his own plan," Allison said.

The smell of baked squash was beginning to smother the mess. Kristian, the boyish videographer from Vancouver, spoke up. "What is the primary purpose?" he asked. "Is it to foul the propeller, get between a ship and the whales? What's it going to be?"

"We're still talking about it," Alex said. "We don't know if we're going to find a factory ship, killers taking the whale back, or what. We wanna be prepared for all of those different scenarios."

"Paul does have his own plan," Allison said. "All these others are secondary."

"As long as we kick some Jap ass, I'm happy," Wessel announced.

"So you're going to head up the special ops toys?" Jeff said.

"That's right."

Alex said, "We do have a lot of steel cable. Maybe their propeller is surrounded by a circle."

"Maybe the Zodiac needs to carry something that will reflect radar. If I'm going to go in the Zodiac with a camera I'd feel much more safe," Kristian said.

"The last time we were here the visibility was fifty miles; we could see anything."

Allison said, "Every Zodiac will have a GPS programmed, and you can reverse it. Chris will train them."

Casson said, "Little by little we have food going off. Can somebody show me how to prepare the pie cannon?"

"It shot about 100 feet. It's more of a gimmick. It's not gonna do much. If Trevor's OK with that we can test the water cannons. What we should do is have the Zodiac try to board and the water cannons hit them. The Japanese—their water cannons are much more powerful than ours. What about wearing a wet suit underneath?"

"If you fall into the water with that wet suit and a Mustang suit you'll be fine."

Steve said, "Yesterday, we didn't know if you had both teams wearing anything, whatever the battle plan was—"

Alex cut him short. "There wasn't a battle plan. Just send the Zodiac out with three people."

Wessel said, "I think if no media are on board, should only be three people. Those Zodiacs aren't big."

"Four is good," Alex said. "You can have one on communications, two for light handling, and if you lose one you have a spare."

Laughter.

The meeting broke up. I went out on the main deck to catch some air just before lunch. Prop foulers, can openers, and pie cannons. In Monty Python they catapulted a cow over the castle wall, which was definitely not vegan. Just as I got to the rail I heard the whale alarm sound. We were still following the ice cliff, maybe 200 feet on our starboard. Up ahead I saw them, two blows, a billow of mist trailing downwind and mingling with the fog. They swam right at the *Farley* and passed within fifty feet. My God, they were big. The smooth glossy back, gray black with one small fin, seemed to go on and on. I heard the gasp of a blowhole, like an air brake. They were two adult fin whales. They might have measured seventy-five feet in length and weighed eighty tons—the largest mammals on earth aside from the blue whale. They were unconcerned with the ship, perhaps curious, and they glided by like a distillation of wild goodwill. An eighty-ton, warm-blooded nod in an ice-filled sea. They were endangered. The IWC estimates that there were 85,200 fin whales left in the southern hemisphere in 1979, but other scientists claim that today there could be as few as 2,000 to 5,000. These were the other species the Japanese were hunting this antarctic summer. They planned to kill ten. Next season they would kill ten more, then fifty every year after that.

By the 1970s, the fin whale had been driven to near extinction. It now numbered about 10 percent of its pre-whaling population. A recent study of pre-whaling populations using DNA analysis, published in *Science* by geneticists from Stanford and Harvard, stated, "In light of our findings, current populations of humpback or fin whales are far from harvestable." The authors added, "Minke whales are closer to genetically defined population limits, and hunting decisions regarding them must be based on other data."

Big Marc was beside me at the rail. He winked and grinned and clapped me on the back. Then he knelt back down one knee beside his pride and joy on the deck. The can opener was looking more and more vicious.

"Yah, we make it ready," he said, picking up his face guard. "Putting it outside will be difficult. Put it where ship is strongest, facing forward. Two and a half meters it is."

"You think we'll get that close?"

"Yah, I hope. Very important to let the bad motherfuckers know there are people who think it's not OK." He pulled down the face mask and lit his torch with a hiss.

■ ■ ■

I went to the stern. I untied the broken wood chair that I'd lashed to a pipe, and retied it with a longer leash. I sat in the cold breeze and felt the push of the deck surging and yawing, heard the boom as the swells sprayed up against the ice wall and watched two Cape pigeons fly over the wake in a tight pair. They were the size of a laughing gull, black with white markings along the tops of the wings. A little while later a wandering albatross swooped across the wake and hung right over the stern, balancing on wings eleven feet across. The longest wings of any flying bird on earth. He teetered and held his position. Dennis had told me that some albatross can fly the girth of the earth without rest.

I went up on the bridge. All the officers were there. It was Alex's watch, but Watson sat in the captain's chair. He was fuming, but nobody who hadn't been observing him for a couple of weeks would know. It was revealed in the energy with which he was telling a story about the Confederate raider the *Shenandoah*.

"Many of the great whales would be extinct if it wasn't for the *Shenandoah*. She went after and decimated the Yankee whaling fleet in the Pacific. Destroyed forty whaling ships. A good role model, too, because her captain, James Waddell, did it without taking a single life."

He motioned at the iceberg out the window. "More damn fresh water in that thing than in Lake Ontario." He finished his story and swung out of the chair. "Got an e-mail from Shane the Rat," he said to me and went back to the radio room. Shane, the Greenpeace expedition leader. Watson handed me the letter.

"Dear Paul,

Thanks for your note. We are also currently patrolling in search of the fleet, but thus far have only been successful in sighting a spotter vessel. When we find the fleet, we will advise the location to all parties through a press release." He went on to say that Greenpeace did not wish to enter into a formal cooperative relationship with Sea Shepherd. "As you noted," he continued, " . . . there have been differences and disputes between our organizations over the years. I do not wish to comment on those, since there is little value in that at a time when we should be focused on the whalers. However, no relationship can be repaired with a flick of a switch. Building trust takes time."

He concluded by saying that it should be possible for the two organizations to pursue their respective missions without interfering with each other. "That is our preference. If we can find peaceful coexistence on this expedition, that may open further opportunities at a later date, but let's take a single step at a time."

I could see why Watson was livid. He had founded Greenpeace and now in a bid to cooperate was being shrugged off by the young sons with a patronizing and sanctimonious note.

"Why wouldn't they want to cooperate?" I asked. "The real reason?"

"Because they know that when we show up we'll actually do something. Stop the killing. Make them look bad." He let out an uncharacteristic exhale of frustration. "You know, I don't think they want to stop the whaling. They make too much money. They raised $135 million off this campaign."

Watson sat down in his swivel chair in the tiny radio room and banged out the following response. The boat pitched and his fingers flew over the keyboard without pause.

Dear Shane,

If your focus is to stop the whaling you would cooperate with us. . . .

Let me tell you something Shane, I have a crew of 43 dedicated crew members out here. They are all volunteers. They have come from 12 nations, spending their own money to join the ship for the purpose of stopping the Japanese whalers. You on the other hand are paid. You have no commitment and no dedication because if you did, you would not allow politics and corporate bullshit to get in the way of the common goal we both should have and that is to stop the slaughter of these incredibly intelligent and wonderful creatures. Whales are being killed in the sanctuary you say, and you don't want cooperation to detract from that central issue. So why have you made it a distraction when a simple agreement to work together would've made both our campaigns more effective and efficient? I know that you have the money to waste and we do not, but it seems to me that if you did not want to make this an issue, you would've simply agreed to work cooperatively. When has cooperation ever been a distraction? When has cooperation ever been a negative approach? Let's face it, the real problem is entrenched bureaucracy and anal retentive career environmentalists who are more concerned for job security than for actually saving the whales.

I give up, Shane. After decades of trying to work cooperatively with Greenpeace, I give up. You people are just simply unbelievable, so full of yourself and so cocksure and conceited that your way is the only way. . . .

Captain Paul Watson

He hit send.

In the middle of the afternoon we cleared the end of the ice island and turned due east along its northern edge. We were about 300 miles east of the south magnetic pole when we made the turn. The little compass on my watchband swung around as if it were drunk. I asked Alex why we were heading farther east. He said that Watson wanted to check out the area around the Balleny Islands off the Oates Coast of Victoria Land. He still felt that the whalers might be working to the east, staying away from Greenpeace.

"Why not farther south?"

"The whalers wouldn't be whaling south of where we are. Too much ice. Plenty of whales around—why go there? I don't think they're around here. Just a feeling. From all the footage I've seen of the whalers, they're usually around pancake ice, just frozen ice, not this glacial ice."

On the main deck, Chris Price and Pawel were teaching the Zodiac crews how to use the GPS-radios.

"The default channel is channel 10," Gedden said.

"Keep all communications as concise as possible," said DMZ Steve.

"What about deciding on a code?" said Allison the warrior.

"Let's learn how to set a waypoint and go to it and then we can get a bit more complicated," pleaded Pawel. You could not fault the Shepherds for their ardor. The training session broke down when another whale, this time a humpback, blew off port.

After dinner, on the way to my berth, I passed Gedden leaning into Colin's cabin in heated discussion. Colin huddled over a sheet of paper on the workbench.

Colin's voice: " . . . go after the shaft and the rudder and propeller. Like a flail. It's a safety thing. I think if you use cable and it gets caught in the propeller it will act like a big wire saw."

Gedden said, "My concern is that if you use longline as a feeder line it'll cut right through."

They went quiet when I stopped. "Prop foulers?"

They looked at each other. Gedden nodded. "Ops props," he said.

"Explain."

"The best thing we could hope for," Gedden said, "is a steel cable wrapped around the propeller shaft, potentially breaking some of their seals so it goes inoperable. Probably the most likely method is coming up on the rear quarter of the vessel and feeding a smaller line into the prop wash and having the smaller line suck in the cable. Attach it all to a buoy so if it kicks free we can pick it up and use it again."

I asked if it had ever worked before.

He pushed his thick black-framed glasses up on his nose and

laughed his machine-gun burst. "Various governments throughout the world have tried it on us—mainly the Norwegian navy, the Faroe Islanders. They were dropping nets in front of the bow in hopes the prop would suck it up and get fucked by it. Didn't work. We ran right over them, chopped them up. That's why we want to use cable."

Gedden explained that they might also try using oil drums as buoys—the Zodiac could run in front of the ship and drop the line across the bow. He and Colin continued discussing the relative merits of feeding cable into the prop or just chucking the whole thing overboard.

"What if they have prop guards, like Allison said?"

"But what's a prop guard?" Gedden said. "A tube around the prop. Their actual function is nozzles—sort of like streamlines your propwash or something so you get more pushing power. Up to about ten knots it creates more power, but higher than that and it becomes a deficit. If you get a mooring line or something caught between the prop and nozzle you can actually jam it, the prop, so if they don't get the engine shut down right away you can do some major damage internally. The propeller's not going to cut the cable. It seems if you're in the rear of the vessel and you drop a lead line in, it should in theory suck it right up. That's the fear that if you fall off the vessel you get sucked up in the propeller."

Gedden said he was committed to running a Jet Ski. "It's horribly exciting," he said.

"Well, I've got my navigator. It's Steve," the good soldier Colin said in conclusion. He was sitting at his tiny desk, and I noticed he had a 3-D commando war game on his laptop.

I followed Gedden up the stairs to the bosun's locker. Inde was in there, sitting on a low stool, splicing half-inch steel cable into an eye. Gedden picked up a heavy stainless shackle from the workbench.

"Uh-oh, I hope they're not planning to use a shackle like this. Look at this, Inde—we can't let them use this shackle. That's a horrible demise for a really nice shackle."

He tossed it back into a bin against the wall. He began to look around the locker, his design gears turning. The other night he had showed me his "Octopus," the rope system he'd invented to support a hanging platform with lines connected to about five acres' worth of tall trees. The platform hung high in the old-growth Douglas fir and the activist camped on it, precariously. Should loggers attempt to cut down any of the trees within the Octopus, they risked severing one of the lines and sending the Earthfirster plummeting to her death.

"There's some butyric acid on the ship," Gedden said finally. "We need better delivery methods."

"What's butyric acid?"

"It's not like debilitating, it just really fucking stinks."

"It's derived from rotting cheese milk," Inde said.

"Nothing stinks as bad as rotten tofu." Gedden fired off a volley of laughter. "A five-gallon bucket of tofu that had been meant to go out to a forest action and had got left in the house all summer, because the lame-ass forest kids don't take care of the house."

"It was terrible. The War Pony was $400. That's our car. The windows didn't go down. The front seat isn't really attached to the floor. We packed all this stuff in it and we tried to ride to the dump, with the bucket. I had my head stuck out the little vent in the window. I was gagging; he was fine. It was probably the foulest stink I ever smelled in my life." She nodded toward Gedden.

"The bottom lip caught when I was pulling out the bucket and it spilled all over the back of my car. As soon as it spilled this woman twenty feet away dry-heaved. I went to the lawn clippings area and dumped the barrel, got lawn clippings all over the car to absorb it. The fourth floor of the house was complaining of the smell from the alley. Butyric acid is like that. Think of some gourmet cheese like Limburger or some shit like that and magnify it by like 100."

"If you break a bottle of it on the deck of a ship they can't be there."

"You know how all the Mustang suits smell like barf? That's not barf." She smiled.

Much later, after the *Farley* had cleared the north side of the iceberg, which turned out to be over forty miles long, and turned southeast, I went into the mess and found Wessel and Casson in intense conversation.

Wessel said, "How do you know we're not going to sink a ship full of Japanese? Fouling a prop can potentially sink a ship. What if a bad sea comes up? You foul the mother ship's prop, and in five days time a big storm comes up and that ship is left—" He tipped back his beer. "Endangers lives," Wessel said more emphatically. "If we could, we would foul all those ships' props."

"I don't know," Casson said. "It's Paul's call. He's not telling anyone."

"He would sink every goddamn ship."

"No, Paul would never sink a ship at sea."

"Yeah, all but one," interjected Shane, a young hippie from Tasmania who was listening from the next booth. "So it could rescue the others."

"That's impossible," Wessel said. "There's 128 people on the mother ship. Twenty on each of the other ships. All I'm saying is it'll even be a squeeze with all the people on the mother ship. All I'm saying is, you're still endangering people's lives. Fouling props in Antarctica is no funny business. If we foul one of theirs, what's to keep them from coming to us and fouling ours?"

"Hopefully it would never come to that. We've never harmed anyone," said Casson gravely.

"You go full steam ahead with that thing sticking out, you can't hurt anyone?" Wessel was talking about the can opener.

Casson cricked his neck, as if the prospect of using the blade was making him tense. "That's why I'm not too happy with that on the ship."

"You have no way of knowing whether fouling a prop will kill people as well. If something like that happens, you can't distance yourself from this organization."

"That depends on Paul's actions. If we go out there and disable a ship and put people in danger, then we need to get them off that ship ourselves. And if Paul doesn't do that, damn right I distance myself. I don't think Paul would ever hurt anybody. That's why I'm here."

"All I'm trying to say is that in this kind of action, there's a possibility of people getting hurt." The freckled, snub-nosed Zodiac driver was calling it like it was; he wasn't shying away from the action, but he was being realistic about possible consequences.

One night in the green room, I had seen a video of one of Watson's actions in the Pacific, ramming the stern of a Japanese drift-netter, and it was hard to believe no one had gotten hurt. The impact of the two vessels almost tore off the power block at the stern. The Japanese crew—most of them froze and watched the oncoming trawler with shocked disbelief, the way you might react if Godzilla were about to step on your house.

I wandered up to the bridge. Watson was in the radio room, writing some late diatribes. Pawel was in there, too, hunched intently over a laptop. He was the only one on board who knew how to hook up to the web through the ABC Australia satellite gear, and he was surfing for news of Greenpeace and possible contact with the Japanese.

"Here's something," he said. "One of the Japanese crew has appendicitis and needs a medevac to Hobart. It says that one of the killer boats is bringing him back to Tasmania for treatment."

"Huh," Watson cocked his head. "That would put us in the right vicinity. We're almost due south of Tasmania now. You wouldn't think the fleet would want to get too far off. You'd think they'd want to rendezvous later."

He went back to the keyboard. I could see the monitor. The piece he was writing was titled "Australia Cedes Territory to Japan."

I went down the outside steps off the starboard bridge wing, taking the outdoor route to bed. It was after one a.m. As I passed the bridge I noticed that our bearing was 155 degrees SSE. We were apparently heading for the Balleny Islands, about 220 miles off. I'd heard a rumor that Watson wanted to land on one of them.

The wind was up and so were the swells. It looked as though we were heading for a massive barrier of low ice that completely blocked the horizon. The light was flat, gray, dusky. Small chunks of ice floated by, and off to the south was a small island, a single pointed mountain of white. The iceberg distilled and intensified the low light. The barrier was actually a low strip of lighter sky lying along the horizon. This twilight world was so deceiving. Hammarstedt had told me tonight that the ice sheet at this meridian extends about 150 miles from land. We were about 180 miles off now.

Trevor was at the bottom of the steps in his blue coveralls, jacketless, smoking.

"You think the weather's changing?"

"I never pay attention to the weatherman when I'm on a stinkpot. What are you going to do? You're not fast enough to outrun it. Do the old-fashioned thing and head into it and pray like hell."

The swells wrinkled in the fresh wind.

"You think we can catch them?"

"We've got the can opener from hell out there. I can do a few things to tweak the engine to get more power out of her. I can get an extra one and a half knots out of her. Probably two. I can get four, but that would be the end of the engine. It's Paul's—it's up to him. I'd gladly destroy the engine to destroy a whaling ship."

He had a lot of his uncle in him. He cupped his hands and lit another cigarette. He said he was just thinking of his wife back home in Canada. He said she was pregnant.

"You know what it is? Your baby?"

"Girl. We've already got a name. Sealoa."

"Sealoa? It's pretty."

"Yeah, it's Sea, plus Length Over All."

12

The Whale Spoke to Justin

That night, or morning, two humpback whales swam directly at the ship while I was sleeping. The crew on watch said they dived and passed under. Just under. They surfaced and blew six feet from the low main deck. Justin said they raised their heads out of the water and opened their mouths and made an unearthly sound.

"What kind of sound?" I asked him at breakfast.

"I don't know, dude. Like nothing I ever heard. It raised goose bumps. I'm not a religious kind of guy but it was mystical."

I was drinking coffee and eating oatmeal in the crowded mess, and it gave me a shiver too. The two whales had been to starboard a few meters from my sleeping head. Justin thought they were speaking to the *Farley*. Given their proven ability to communicate across thousands of miles in what scientists are beginning to discern as a syntax, maybe it's not too much of a stretch.

Geert had mentioned recent studies showing that humpbacks make 622 social sounds. New research has also found spindle neurons in the brains of humpbacks, fins, orcas, and sperm whales. Previously thought to be unique to humans, these specialized cells are located in areas of the human brain associated with social organization, empathy, speech, and intuition. They are thought to process emotion and are the cells responsible for feelings of love, grief, and suffering. The researchers found that the whales' spindle neurons reside in the same area of the brain as those in humans, that they have existed in whales' brains much longer than in ours, and that whales have proportionally three times more of the cells than we do. Not only are whale brains larger than ours; they have four lobes to our three, and have more convolutions in the neocortex.

As reported by NewScientist.com, one of the researchers, Patrick Hof of the Mount Sinai School of Medicine in New York, said, "It is absolutely clear to me that these are extremely intelligent animals. Their potential for high-level brain function, clearly demonstrated already at the behavioural level, is confirmed by the existence of neuronal types once thought unique to humans and our closest relatives." Hof added, "Dolphins communicate through huge song repertoires, recognize their own songs, and make up new ones. They also form coalitions to plan hunting strategies, teach these to younger individuals, and have evolved social networks similar to those of apes and humans."

That whales are intelligent, are self-aware, and have a capacity for suffering and grief was not in doubt. Why wouldn't Justin's humpback call out to the *Farley* in a mystical way?

Humpbacks seemed to be the most charismatic of the great whales, if only because of their exposure to humans. They were the staple of a worldwide whale watching industry that brought in over $1 billion every year. They had the long lateral fins, sometimes as expressive as arms, and the love of the leaping breach made famous in commercials for Pacific Life. They were also to be on the Japanese menu. In two years, JARPAII, the Japanese "research plan," called for killing fifty of the endangered humpbacks annually. Some estimates put their worldwide population at only 18,000, and some biologists thought that figure much too generous.

The new statistical DNA research on the whales, which can remarkably and accurately assess ancient population numbers and their dispersal patterns, has raised estimates of the humpback's pre-whaling numbers to as many as 1.5 million. Imagine 1.5 million humpbacks patrolling the ancient seas, calling across oceans where only the songs of cetaceans echoed in the deep pelagic blue. There were no engine sounds then, and the water was as clear as krill and plankton would allow. The whales called over great uninterrupted distances in complex syntax while human ancestors were still jumping around in trees. This was the magic of whales, that they had expressed loyalty, grief, gratitude—all well-documented among present-day humpbacks and other cetaceans—and had called each other by name and referred to a third party by name, long before we were even a seed of an apple in God's eye. And they had done it all for millions of years and had swum the oceans in peace. They had left the sea unpolluted, mostly quiet, the reefs teeming; the shores, the mangroves rich, protective; the fish in their schooling numbers as prolific as the stars that wheeled above. They had loved the ocean, if love is a deep attention in which one does no harm. They had perceived it, attended the greens of the reefs and the blues of the deep and all its creatures and passed on, generation to generation. They had not turned on each other with wholesale vengeance and bloodlust, or massacred another species off the face of the waters because they could.

The ocean they swam in now had changed. Old drift nets called ghost nets, thousands of miles of them, abandoned, drifted in every sea; the whales, all the species, could not detect them until it was too late, and then became tangled and thrashed and died by the thousands. Other fishing gear did the same—lines of lobster traps, longlines, abandoned seines. Ships, the sound of engines and props, turned the great currents into a cacophony through which the old distant whale songs were mangled and lost. Low-frequency active sonar now being used by the U.S. Navy, one of the loudest sound systems devised by man, emitted sonic booms that ruptured delicate hearing mechanisms, caused internal hemorrhage, and destroyed cetacean navigation systems so that whole pods washed up disoriented on beaches in the Caribbean and in the Pacific, bleeding from their ears. Overfishing dramatically depleted fish stocks and dug deeper and deeper down the food chain as top predators were wiped out, so that for cetaceans, food was harder and harder to find. Pollution concentrated toxins in their flesh to such a degree that many environmental organizations claim that the eating of whale meat is patently unsafe.

We were now in a land of ice. Where the Japanese were, where we moved, depended much on what the ice would allow.

After breakfast, just at 0900, we heard the wave-washed silence of the *Farley*'s engines shutting down. We drifted. We were a few hundred yards off a flat-topped iceberg the size of an aircraft carrier. The threatening winds of the night had died down, and the sea was calm, rolling easily, wrinkled with a light breeze that blew snowflakes around. The sky was overcast but bright, luminous; and a darker storm front was moving in.

It was time to get the chopper up while we could. We were still about 150 miles from the Balleny Islands, off to the southeast, and Watson wanted Aultman to scout ahead as far as he could and find an opening if there was one.

Watson said, "I'm quite confident that if we go all the way to 175 degrees that way [east], and then come back this way we'll run into them. Last time there were only three legal Patagonia toothfish boats in Antarctica and we ran into two of them."

But the last time he was here, in 2002, he had completely missed the six large ships of the whaling fleet.

With the engines off, it was easy to hear all the other activity on the *Farley*. She hummed with a renewed vigor. In the bosun's locker the grinder whined and smoked. DMZ Steve was at the vise hacksawing a piece of four-foot angle iron for a prop fouler.

He stopped and lifted a smudged sheet of paper off the workbench—a diagram of a pointed cross with the scrawled words "killer steel."

Some Twisted Sister hard rock blasted next door in the coring lab, another workshop next to the bosun's locker—"*I'm not going to take it!*"—where Luke was cutting more angle iron, Justin wrapped a steel cable splice with electric tape, and Jeff was bent over a tangle of cables in the corner.

From all over the boat, loud music, hammering steel, grinding, drilling, tapping.

Trevor and Willie were out at the forward end of the main deck, starboard side, helping Marc lower the can opener into position on the outside of the hull. Marc had on a harness and was roped up. Willie belayed him. Trevor worked the knuckle boom crane that lowered the blade slowly. Once in place, Marc dangled and began to weld.

I went up to the bridge. A very serious folksinger was vibrating the woofers of the bridge stereo with her feelings.

"And what in God's name is this?" I called to Geert, who was drawing at the chart table two feet away.

"What music is this?" Geert repeated loudly to the bridge. He turned and raised an eyebrow. "Shit music."

"It's the captain's," Alex said through the doorway. "I don't know. Let's keep it that way."

I found the jewel case—Connie Dover, Celtic.

I looked at the GPS over the charts. We were at 65°41'S, 155°55'E, eighty-five miles east of the Virik Bank and about fifty miles north of the Antarctic Circle. The Krylov Peninsula was 200 miles to the south.

We were facing almost due south. The sun sat in the southeast, a brighter cloud bank, about thirty degrees off the horizon.

At 1031 the crew lowered and launched Alex on a Jet Ski. They also put one Zodiac, the port side boat, into the water. Kalifi worked the davit.

At 1040 I was scanning the horizon from the main deck with my binoculars and my heart began to thud. I saw a distinct black shape like a ship.

"Hey!" I called up to the bridge. "Is that a ship?"

"It's probably ice," Watson yelled down from his window, very calm. "The helicopter's going out in ten minutes; he'll check it out."

Alex headed in that direction, south, on the Jet Ski, and halfway to it he stopped. His voice came over the bridge on channel 16. "Home-base, this is whale party, over." On the radar I could see that he was only a mile off. Distances were completely deceiving. What I thought was a ship was probably a crooked finger of backlit dark ice about two miles away and the size of a refrigerator.

Alex told Hammarstedt that the engine of the Jet Ski had failed. Hammarstedt sent the Zodiac to tow him back. Hammarstedt said into the mic, "Alex, for future reference use only channel 10." He explained to me that the ship's marine radio carries at least 100 miles.

Lincoln, watching on the bridge, said, "Whenever we send a Jet Ski out there's a breakdown."

Producer Ron, from his armchair, added drily, "The other thing is, they have no sense of direction. I've seen it on a couple of occasions. They take off and, man—"

It seemed like just another morning of exercises on the *Farley*. Except for the chopper. Aultman was all business on the heli deck. He had his engine blaring, and his deck crew circled around him while he went over the hand signals. The crew would be crucial to his safety and his life—guiding him in, centering him, grabbing and strapping the pontoons down fast before the bird could pitch off into the sea. All week they'd been training together—pumping fuel with a hand filter pump out of the avgas drums, tweaking the chopper's tie down system, adding blocks of wood under the engine, repairing a cracked fuel pump, hosing down the machine with precious fresh water when the ship plunged through spray. Theirs seemed to be the one unit on the ship that had been practicing constantly.

At 1121 Aultman and Pawel strapped themselves into the two seats. They had taken the doors off the bubble for better filming. Pawel was in his elite night-colored dry suit and special ops neoprene boots, excited, hanging out the door with his video camera. On Aultman's green helmet was a decal that read, "I'm training to be a cage fighter," next to a picture of a skinny nerd with huge glasses. It was the geek brother in the film *Napoleon Dynamite*.

At 1125 the rotor started turning, and the deck crew crouched and unhooked the tie-downs. At 1129, with the *Farley* in a very light roll, the rpm of the blades kicked into a higher key, the chopper lightened, relieving the squashed pontoons of its weight, and then it lifted, hovering a foot off the deck. Aultman held it there, getting used to being airborne again, and the craft hovered like a hummingbird. And then patiently, carefully, with tight reins, he backed the bird off the ship as he climbed. Thirty feet up he tilted it forward and accelerated, and then he was circling the ship like a wild hawk, twice, gyring upward, waving, and he broke away off to the southeast and out of sight.

Meanwhile, Alex was lifted back onto the main deck. The Jet Ski had a hole in it. They went to start the other one and found the electrical box full of soupy water. After Gedden's fiasco in Hobart bay, they'd towed it back and left it as is in its cradle, unrinsed, undrained. They had remembered to hook the battery up to a pulse charger, which, with the water, fried all the electrical parts. Alex, Trevor, and Price spent much of the afternoon bent over the engine, replacing the burned pieces.

The chopper touched back down at 1244, landing flawlessly. The news wasn't good for a trip to the islands. To the south and east was ice. Aultman left the engine idling and jumped to the deck. Casson handed him a cup of hot coffee. Pawel was lit up but moved stiffly, as if his ligaments had turned to ice. "The weather was shady," Aultman said. "We were dodging that stuff the whole time." He pointed to the southeast at a dark cloud bank shedding snow. "We'd come around; it'd open up to thirty miles, then close to ten. About ten miles ahead the clouds closed right down to the water." That would be his nightmare, being overcome by storm, with no visibility. Like a Zodiac. The ship would prove hard enough to find on a clear day.

"What did you see?"

He shook his head. "Nothing. Ice." He went into the chart room to debrief the captain. The two bent over the chart.

"I'd say we flew twenty miles from the ship. The weather is not really conducive to surveillance. You can see nothing over here. Well, there's a lot of ocean over here." He pointed to the west of us. "The ceiling is 1,500 feet, pretty much steady. It's really a low-hanging day. I would say probably at times I was able to see sixty miles from the ship in clear weather—over here, where we made the turn back west, it was particularly clear; could probably see forty miles. Anyway, that's the route that we flew." He drew a rough rectangle. "The ice extends like this." He drew a line to the south of us, and one to the east, which ran northeast.

He went back outside to attend to his steed. There was a fast drip from the underside of the engine puddling on the rubber of the deck.

He raised his hands, "Oh, man!"

He crawled underneath and swiped it with a finger. Fuel and water. He stood contemplating the leak.

Aultman shut down the engine and the leak stopped. He nodded to himself. "What it is, you force-feed fuel down into the engine to purge the engine—floods the throttle. That's the release valve that spit fuel." He was learning as he went. He said that on the way back he landed on the big flat-top iceberg next to us.

"I got that fantasy right out of the way. Good, solid surface. You don't fool around with icebergs. The wind hits the wall and goes straight up. We got some turbulence. Also, on the lee side of the iceberg, if you get close enough it's perfectly calm. We went around the back side and got five feet over the water and twenty feet from the berg."

Down on the main deck, the can opener was in place and Marc was perched on his baby over the water, welding on more supports. Watson said, looking down at it from the bridge, "If you sideswipe something it'll probably crush that thing like a piece of paper." But it looked vicious. Maybe that was the point.

In the afternoon Pawel and his assistant, the Quiet Australian, George, took their re-breathers and a Zodiac and went diving in the ice. Just to the south of us was a thick mosaic of deteriorated berg and glacial ice. The light snow thickened. Dry flakes blew and eddied on the faintest breeze. They fell out of the sky like dark feathers.

After dinner, the J. Crew and Casson, Ron, Watson, and I played poker. Price was telling us another grisly story, whether we wanted to hear it or not.

"Want me to tell you a story?" he began.

"Not really, Chris," tough Jeff said. "We're trying to have a game here."

"No, really. Listen—the Mormons massacred all those people in Arkansas. They have a marker in a field, but it's not there." He nodded, emphatically. "Go to another field and you could dig up all those bodies! I could show you where . . ." He pointed at an open road atlas.

When Price left, Jeff shook his head over his cards. "I'll tell you what, if I heard they found forty-eight bodies in that guy's backyard I'd feel a lot of things, but surprise would not be one of them."

Later that night, I saw an announcement on the eraser board:

Refresher course on CPR and management of critically injured. I need at least five crew for a resuscitation team—Kristy.

Lincoln had the ship pushing north of east at seventy-six degrees. Half an hour later he was forced by the ice edge farther north to sixty-eight degrees. By midnight we were heading not far off north at twenty-one degrees. There was ice everywhere. A line of it like a frozen coast off to our starboard, chunks and rafts and growlers littering the water. Watson said, "We have to head around the ice—can't go through it. At least we know they're not in it."

After breakfast, a bunch of us were on the bridge, and Watson handed around an article from the New York Times that was e-mailed to him by a friend. "Counterterrorism agents at the Federal Bureau of Investigation have conducted numerous surveillance and intelligence-gathering operations that involved, at least indirectly, groups active in causes as diverse as the environment, animal cruelty and poverty relief, newly disclosed agency records show. . . .

"The FBI document indicates that agents in Indianapolis planned to conduct surveillance as part of a 'Vegan Community Project.' . . . The latest batch of documents, parts of which the ACLU plans to release publicly on Tuesday, totals more than 2,300 pages and centers on references in internal files to a handful of groups including PETA, the environmental group Greenpeace and the Catholic Workers group, which promotes antipoverty efforts and social causes. . . .

The documents indicate that in some cases, the FBI has used employees, interns and other confidential informants within groups like PETA and Greenpeace to develop leads on potential criminal activity and has downloaded material from the groups' Web sites, in addition to monitoring their protests.

In the case of Greenpeace, which is known for highly publicized acts of civil disobedience like the boarding of cargo ships to unfurl protest banners, the files indicate that the FBI investigated possible financial ties between its members and militant groups like the Earth Liberation Front and the Animal Liberation Front.

These networks, which have no declared leaders and are only loosely organized, have been described by the FBI in congressional testimony as 'extremist special interest groups' whose cells engage in violent or other illegal acts, making them 'a serious domestic terrorist threat.'"

Watson thought it was amusing that Greenpeace was on the FBI list of ecoterrorists but Sea Shepherd was not. I didn't think not being mentioned in the article had anything to do with the FBI's list. I had already read the congressional testimony mentioned in the article, and Sea Shepherd was pointedly acknowledged. On February 12, 2002, James F. Jarboe, the bureau's Domestic Terrorism section chief, testified before the House Resources Subcommittee on Forests and Forest Health on "The Threat of Eco-Terrorism." He said:

> Since 1977, when disaffected members of the ecological preservation group Greenpeace formed the Sea Shepherd Conservation Society and attacked commercial fishing operations by cutting drift nets, acts of "ecoterrorism" have occurred around the globe. The FBI defines ecoterrorism as the use or threatened use of violence of a criminal nature against innocent

victims or property by an environmentally oriented, subnational group for environmental-political reasons, where aimed at an audience beyond the target, often of a symbolic nature.

He went on to talk about Earthfirst!'s crossing the line into criminal activities when its members began spiking trees to sabotage logging operations.

Moreover, several members of the crew, including Allison, had had run-ins with the Justice Department for possible association with ALF, ELF, or both. I thought it was interesting that Jarboe had mentioned tree spiking in the same breath as domestic terrorism. Just the other night at dinner, Watson had bragged to me that he had invented the technique in Vancouver in 1983.

"Well, actually, we discovered the Wobblies had been doing it back in the 1920s, to sabotage sawmills." He smiled. "The North Vancouver Garden and Arbor Club. That was us. People say, 'You openly committed acts of ecoterrorism by spiking trees.' I say, 'No, I did it legally—every tree I ever spiked was legal. We did it before the laws were created.'"

There was one other note of interest in Jarboe's testimony. As an example of the bureau's success in the apprehending and subsequent conviction of ecoterrorists, he mentioned Ron Coronado:

Rodney Adam Coronado was convicted for his role in the February 2, 1992, arson at an animal research laboratory on the campus of Michigan State University. Damage estimates, according to public sources, approached $200,000 and included the destruction of research records. On July 3, 1995, Coronado pled guilty for his role in the arson and was sentenced to 57 months in federal prison, three years probation, and restitution of more than $2 million.

Coronado was one of the two men Watson had sent to Iceland to sink the two whaling ships in the winter of 1986.

■ ■ ■

It is December 21, summer solstice in Antarctica, the longest day of the year. My GPS claims the sun sets for an hour and a half before it rises at one-eighteen a.m., but we have not seen the sun for days. The only darkness the sky allows now is ominous storm fronts, almost as black as the smoke from forest fires; they move along the horizon covering the sea with what must be blizzards, but so far they have left us alone. We are zigzagging north in a Monument Valley of ice, trying to get around the frozen edge lying to the south and east. To the north and west is slate-gray open water scattered with great flat-topped monoliths like mesas. We move through them dwarfed like a defile of cavalry as it enters the silence of cathedral stone. They could be stone, these floating islands. Shadowed blue by distance or blazing strangely with a captured light. Strange because the lowering, rolling overcast does not betray it. The sun never breaks through, and the cliffs of ice miles away seem to emanate a cold white fire.

Watson wants to send the chopper up again to look for a passage to the Balleny Islands. He has started to call the Japanese ship we seek the *Death Star*.

At 1029 we cut the engines again and drift. The whole landscape, cliff islands and all, moves with us, two knots westward, on the antarctic coastal current. Aultman takes off an hour later, with Ron this time. While he is gone, Geert, Marc, I, and a few others who will be on the *Farley* during any action take a refresher course in CPR and critical care with little Kristy in the green room. She shows us how to use the oxygen tank and mask. She straps Mandy into a folding stretcher, and Marc and Geert hoist her out of a simulated sea as if she were made of balsa wood.

Outside the sky was moving faster than the current, low to the water, and layered with shades of blue-gray that changed mood every moment: now infused with brightness, now threatening. Two hours after the chopper took off one of the dark cloud banks that veiled down to the water overcame the ship and shrouded us in thick snow. Every ear out on deck was listening for the throp of the helicopter.

The blizzard passed and the lid lifted and we rocked easily. Trevor and Wessel smoked by the starboard life raft. Trevor said, "We lost one of the pistons on the knuckle boom—the seal blew."

Wessel held his cigarette low, shielded by his leg. "Did it scream? The seal?" He smiled insolently at the chief engineer.

Trevor looked at the South African. He ground his butt against the superstructure. "And you wonder why it's so hard for you guys to get visas to go to other countries." He went through the hatch in the stack and descended back down to his entropy room.

Aultman landed from his second flight at 1340. Low clouds had socked the ship in all around, and it was hairy flying back. His deck crew guided him in perfectly and caught the pontoons fast. Ron climbed out with his big video camera. He said, "Had a great time. We saw two whales, I think fin whales, and no Nips. Flew 1,400 feet when we could, then came down when the visibility went down. We went east for about forty miles, then north. The whale was huge, man. It was fucking big."

Aultman said, "We were 100 feet off the water. The whale was blowing. A smaller whale another 200 yards to the north of it. They were going east, right toward the ice. Visibility was great to the north. We both thought we could see sixty miles."

"I was glassing it," Ron said. "You can see quantum leaps more with binoculars than with the naked eye. The ship looks pretty fucking small coming back, pretty alone."

"As far as I could tell, sixty to seventy miles north of here, to the east, is all ice. And to the south was ice."

Hammarstedt pulled out the *Sailing Directions*. "It says that between the islands and the mainland is almost always impenetrable."

Alex at the chart swept his hand between the coast and the islands. "OK, this is basically a no."

The engines growled to life and we headed east a few miles to the ice edge and then followed it north, looking for a way around.

In the mess the J. Crew were recovering from an hour's hard work with a paint roller. They were capable of extreme concentration at the poker table, but when it came to manual labor they were desultory. Jeff looked rough—he had been e-mailing his wife every day, and his marriage was tenuous—but he was excited.

"You should see the toy I just built. Come on." He dragged me forward to the coring lab. Youth Gone Wild was blasting from the boom box. Amid the clutter of scrap steel and tools Jeff picked up a longline buoy the size of a beach ball, hung all around with coils of steel cable.

"This is called the Jellyfish. If you look at her, she's compact." He motioned with his other hand like the woman on *The Price Is Right*. "Got three tentacles. Each comes out on the side. Each tentacle is seven feet in length. On each tentacle is one or two eight-foot to twelve-foot rolls of seat belt strap placed strategically so as not to interfere with each other. For deployment the Shepherd just has to toss this bad girl into the port side or starboard side—first snip the threads holding the coils.

"Any one of these gets tangled in a prop—let's just say it's like the gay man in the whorehouse. He's not doing anything; he's just going to enjoy the scenery. I field-test it tomorrow. I made it economical in the sense that it can be re-collected if it doesn't find the designated target. Really easy to construct as well. Pending field-test approval I can construct dozens of these in a day."

It was an impressive design.

"I'm thinking of calling her Rebecca after that Alfred Hitchcock movie. *Rebecca* is so intriguing and entangling, I remember when I saw it as a kid I was entangled."

The waiting was wearing on everybody, and the crew channeled their energies wherever they could. Most of the ship retired to the green room to watch the movie *Phone Booth*, which is about a man who is also waiting in a terrible limbo, waiting for the call.

Emily burst into the dark cabin and announced, "Greenpeace found the fleet!"

13

Uninvited Guest

2040 hours

64°07'S, 157°28'E

Watson was leaning over the chart table. He had on his reading specs and he was holding a sharpened pencil and the parallel ruler.

"Apparently Greenpeace located them around here. Somebody in Australia tipped us off on it." He moved the pencil over an area of sea about fifty miles north of the Mertz Glacier tongue on the George V Coast. Far to the south and west of us—311 nautical miles. At nine knots that was almost thirty-six hours. A day and a half away.

The chart room was suddenly full of crew, and Watson cleared it out. But the word went through the ship like an electric current. Everybody knew the coordinates, the distances. I looked out the bridge windows and saw the little nurse Kristy dancing wildly alone up on the forecastle, on the bow, to some music that was in her head, her nest of dreadlocks flying.

Allison said, "We got a phone call. From somebody in Melbourne."

Watson said to his second officer, "Peter, can you set a course for sixty-six degrees south, one hundred forty-six degrees east?" The beanpole shoved his little wire-rimmed glasses up on his nose, smiled his impish smile, and nodded, "Yes, sir." He took the parallel ruler down from its peg and got to work.

"Two hundred forty-three degrees," he called out a minute later. West-southwest. Lincoln was at the helm, and he turned the ship in a tight arc. The *Farley* rolled and then settled into the new heading. The waves, which had been on our port bow, were now quartering onto the stern, pushing the *Farley* south. The engines seemed to thump with a slightly different tone.

Jeff was on the bridge deck. Crew members ebbed back to the ship's nerve center. He said, "We need to build more toys."

Allison said, "Chris Price, we've got to get our thing up and running." They had been lovers once, and she called him by both his names. She was referring to the FIB.

Aultman said, "It certainly shrinks the ocean down. Getting some altitude."

"I'm going to have a really great Christmas!" Emily chimed, her ponytail bobbing in her excitement.

Watson said, "I've got to get more information," and headed for the radio room.

Aultman turned to me, "The way it was explained to me, Greenpeace has satellite data that located the fleet. Who else would have six ships grouped within 100 nautical miles of each other?"

I went over to the chart and found Hammarstedt's little pencil cross and the initials GP, Greenpeace. Every hour, the officer on watch took down the ship's position from the GPS, marked it on the chart, and extended a pencil line from the last position. I followed the ship's track south to the point where we had turned around in the ice pack and fog—where the penguins had stood—and traced the stuttering line back to the north and east where it struggled to find a way east

around the ice. We had been very close to the Japanese without knowing it. From that most southerly point to where they were now, it had
been only about 120 miles, a little over twelve hours away. But without intelligence they might have passed us and we would never have
known.

Hammarstedt was concerned about ice. We were heading for a
point two degrees of latitude farther south. A degree of latitude spans
sixty nautical miles. So that was 120 miles closer to the south pole, and
we were already dealing with ice as obstacle in many forms, from
huge island bergs to littered growlers to broken rafts that packed
together in a tight mosaic to form its own false coast. Hammarstedt
pulled down the *Antarctica Pilot* and read.

"Says here that Watt Bay, just on the other [west] side of the Mertz
Glacier tongue, is ice-free. The coordinates we were given are about
thirty miles north of Mertz glacier tongue. Maybe the ice moves west
with the current. So we should be OK."

It was 2100 hours. Watson was in the radio room firing questions
into the Iridium sat phone.

"That's about 300 miles from where we are now. Have you checked
their website? Have you actually nailed the position down? . . . Uh-
huh." (Listens.) " . . . Speak up; I can't hear you. What is that? I don't
know what you're referring to—Is that because that's where they're
medevacing this guy? From the Japanese vessel? OK, who is it from
Greenpeace that called Paul? What did they say to him? Uh-huh. And
that came from Greenpeace? How does he know what's going on with
the fleet? He was—I don't know. You don't think we should post anything on the website? No, I mean say specifically how we got that
information? Yet. Well, there's no—thank you. OK, we'll just have to
see what we can. I'm just worried they'll get out of the area. If we get
that close, we can use the helicopter. OK—" He hung up.

"Now I'm really pissed at Greenpeace," he said. "They specifically
will not put their position on the website, and they're doing this so
that Sea Shepherd doesn't get the info. It's the first time in history
they haven't posted it."

He leaned back in the chair, and it squeaked and cricked under his weight. "They're medevacing a Jap guy with appendicitis out to Hobart and this contact in Hobart called—got it from the hospital. Called our guy in Melbourne. They're doing it on Christmas day. Don't know why they're waiting. This Shane Rattenbury, the Greenpeace expedition leader, is saying that. So they put out their Zodiac today in front of the fleet and displayed a banner that said, 'Get Out of the Southern Ocean Sanctuary Now.'"

Watson made a face, like "How lame is that?"

"Paul Martin in Melbourne receives a call from someone in Hobart, some guy working with the hospital there. So we got it independently. Paul Martin called Greenpeace to get the position and Shane told him that we couldn't have it." Watson didn't fume like ordinary people. The only way you could tell was that his voice got softer.

"One of their officers apparently calls a meeting on the *Esperanza*, with Shane the Rat, says he's fed up with the bullshit on board—that they won't cooperate with us and they're not doing anything. They're taking pictures and hanging banners." He shook his head.

"By tomorrow afternoon we should be within helicopter range of that position," Watson said.

Just then Pawel stepped into the tiny room and hooked his laptop up to his satellite dish. The first thing he did once he got a connection was pull up the Greenpeace website.

Pawel turned his screen toward Watson. The captain gave it a cursory glance. "That picture's ten years old," he said. But there was a new video as well.

Others filed in to look. Alex was the most interested. The serious first officer studied the video clip with an almost salacious intensity. "Fuck," he muttered to himself, and went back to the bridge. He turned to Jeff. "Did you see what Greenpeace was trying to do with the Zodiac? They were trying to steal the Japanese fucking banners. Like they don't have enough of their own." He checked the radar, then stared out ahead in a kind of reverie. "That middle Greenpeace

Zodiac is beautiful. I love that one. It's bigger, too. Maxed out, it's got a 140. They could get more." He was talking about their outboard motors. He looked down at his own sad Zodiac fleet lashed to the main deck. The biggest, the center one, had an old 115-horsepower. Zodiac envy.

I asked Alex if he was excited. He blinked. "No, I don't get excited. I'm Dutch. We don't have a sense of humor that we know of." He tightened his mouth into a quick smile. "Nah, I'm not excited. Waiting three fucking months for this."

Watson couldn't keep himself away from Greenpeace's website. When I returned to the radio room he was reading its account of the skirmish.

"'We will not harm you or your equipment,'" he read aloud. "'We will use all means necessary to prevent you from killing whales.' What the hell does that mean?" There was a picture of the *Nisshin Maru* looming behind the festively painted *Esperanza*. Watson's finger tapped the arrow key loudly as he scrolled down. "Let's see, Andrew's blog. Good old Andrew. It's titled 'Rough First Day.' Says here he had to run in and get lunch in case there is more action."

"'No whaling, though . . . ' Didn't I just read they were bringing back two minke whales?" He turned the laptop away in disgust. It was what he knew he could do with a ship that fast if he were that close to the Japanese.

An hour later the sat phone in the radio room rang. Watson listened intently; said, "Thanks," and hung up. It was one of the crew members on board the *Esperanza* who was sickened by the savage killing they were witnessing. He was also tired of the bureaucratic stonewalling and knew Watson could stop the whaling. He called in secret, at the risk of losing his position with Greenpeace.

"The position he gave us is thirty miles closer than the one we have," Watson said, going to the chart. "He thinks they're heading north but doesn't see them moving around much." He plotted the new position: 65°55', 146°33'. "Hey, Al," he said to his wife, "Contact Tim Midgley at the home office and get the French station phone number." Sixty miles west of Commonwealth Bay on the Adélie Coast was a French research station at Dumont d'Urville.

Alex said, "If they are coming toward us, then we might see them tomorrow. But if they're going away, it could take us a week. If the French don't see them, then they are either coming this way or stationary."

That made sense.

Watson said, "I asked him how long they were going to hang around. He said three weeks. I was thinking, 'Not if we get there.'" He set the parallel ruler between our current position and theirs and drew a fine pencil line along it. "Two hundred forty-seven degrees," he said to Hammerstedt.

"He also said they had been told that Greenpeace posted their position on the website," Watson said.

Allison narrowed her eyes. "But the Rat said no. So we wouldn't get it. Greenpeace is there; the killers are there—everybody is there but Sea Shepherd. We are the life of the party."

"We are the uninvited guest," Hammarstedt said.

Watson leaned against the chart table. "I went to the IWC meeting in Monaco in 1997. I put on the master's uniform. About 600 people at this reception. Everybody is staring at me. The Japanese, the Icelanders, the Norwegians, 300 people walk out of the reception in protest. McTaggart [then chairman of Greenpeace International] walked out too. John Frizell, the director of Greenpeace UK, comes up, says, 'It's not nice to come to places you're not invited.' I said, 'I thought that's what Greenpeace was all about. Showing up at places you're not invited.'"

Early the next morning, Pawel pulled a story off the Internet, from Melbourne's daily paper. In the vastness of waves and birds and ice, these missives were a concrete reminder of the other players, and of a world beyond the ship and the bounds of the horizon, and we read them avidly, start to finish.

GREENPEACE CLASHES WITH WHALERS

by Andrew Darby, *The Age*, Melbourne, Australia,
22 December 2005.

Greenpeace activists have found and tackled the Japanese whaling fleet in Australian Antarctic waters, opening the most extensive campaign yet to halt the "scientific" hunt.

The fleet was catching and processing minke whales when it was tracked down by Greenpeace ships *Esperanza* and *Arctic Sunrise* yesterday. In an initial skirmish, the *Esperanza* and a catcher ship collided in what Greenpeace expedition leader Shane Rattenbury said was a violation of the rules of the sea by the Japanese.

The encounter opens the first hostilities between the two sides since water cannon aboard the factory ship *Nisshin Maru* forced Greenpeace protesters to retreat in early 2002. It comes as Japan doubles its self-awarded quota of minke whales to 935 and includes 10 fin whales for the first time.

Mr. Rattenbury warned that Greenpeace's 61 campaigners, who have strengthened their defense against the water cannon, were there to stay.

"Our small boats will be doing everything they can to get between the harpoons and the whales," Mr. Rattenbury said.

The whaling fleet was located within 160 km off the Antarctic coast near Commonwealth Bay yesterday.

The location puts the fleet deep inside an Australian whale sanctuary, but a Federal Court judge said earlier this year that an attempt to prosecute the whalers was futile, since Japan does not recognize Australia's Antarctic claim.

For the first time, Greenpeace is devoting two ships to the antiwhaling campaign, and the hard-line group Sea Shepherd is also sailing to confront the fleet with its vessel *Farley Mowat*, raising the prospect of the whaling program being severely disrupted.

Yesterday's collision happened when the Greenpeace ships

steamed in close behind the *Nisshin Maru*, preventing a capture boat, *Kyo Maru No. 1*, from transferring the minke whales it had caught.

"They started attacking us with water cannon and sounding their horns," Mr. Rattenbury said. "Then when we didn't move, *Kyo Maru* turned into *Esperanza* and struck it. It was a touch more than a direct ramming, but a touch nevertheless."

He said that when the *Kyo Maru* turned in towards the *Esperanza* again, the Greenpeace ship backed off.

Federal Environment Minister Ian Campbell last night repeated Australia's call for Japan to respect the International Whaling Commission's condemnation of the hunt. He also confirmed that one of the whaling fleet is due in Hobart on Christmas Eve with a crewman who is understood to have appendicitis.

After breakfast, Watson handed me an e-mail he'd just gotten from the crew member on the *Esperanza*.

Dear Paul,

One more thing. The Japanese whalers DO indeed have firearms on board.

I heard at least 4 shots when they had a whale on the harpoon cable right under their bow. A crewmate of mine said it was a big fuck-off rifle.

Just wanted to update you.

All the best

Now everybody was armed. Except Greenpeace, whose people were "Quakers with attitude," according to Watson. They wouldn't have a big fuck-off rifle, would they? The only actors we hadn't heard from were the Japanese, but word came in the form of a press release a few hours later. It was dated the day before.

For Immediate Release

December 21, 2005

Safety Fears in the Southern Ocean

The organization that conducts Japan's research whaling in the Antarctic said tonight it feared for the safety of its crew and scientists after one of two Greenpeace vessels in the Southern Ocean collided with a Japanese vessel.

The Institute of Cetacean Research (ICR), which carries out Japan's research whaling in the Antarctic, said the actions by these environmental groups were putting at risk the crew and scientists of JARPAII vessels.

"We have told Greenpeace to keep their distance from our research vessels. They are severely compromising the safety of crew and scientists," the ICR Director General, Dr. Hiroshi Hatanaka, said.

The same thing occurred five years ago when in 1999 another collision occurred between Japan's research vessels and a Greenpeace vessel. "Greenpeace and any other group in our vicinity now are acting unlawfully, recklessly, and absolutely irresponsibly. This is beyond legitimate protest."

Japan's research whaling is perfectly legal under all relevant international law and authorized under the International Convention for the Regulation of Whaling.

"Our whale research program provides vital information for the management of Antarctic marine resources, which cannot be gained using only nonlethal methods. The number of whales to be taken in no way threatens these abundant populations," Dr. Hatanaka said.

"Greenpeace, Sea Shepherd, and other antiwhaling organizations have been misleading the public with their antiwhaling campaigns for years. The campaign has no conservation or scientific basis, and is simply a continuation of the misinformation and publicity stunts they've used before for fund-raising purposes. They should move away from the area," Dr. Hatanaka said.

Step away from the sanctuary.

The International Whaling Commission (IWC) Scientific Committee had sent Japan twenty protests over the eighteen years of JARPA. And in all that time of gathering vital information, the Institute of Cetacean Research had published few peer-reviewed scientific papers, and not a single one in the IWC's *Journal of Cetacean Research and Management*. In fact, a review of the Japanese program by the Scientific Committee of the IWC in 1997 declared that the research failed to meet its stated objectives, and that the data were not required for management. The ICR said that nonlethal means could not "gain" the vital information on marine resources. But the World Wildlife Fund had recently published an extensively supported report showing how current nonlethal techniques of gathering skin tissue revealed far more—through DNA and chemical analysis—about population health, sexual maturity, and stock structure than the decades-old methods the Japanese used. The Japanese claimed that one crucial data set, however, can never be garnered through nonlethal means: a whale's age. Which they determine by fixing it—the age—with a harpoon and then examining the ear bone. But a recent report in *Science* by a team from Australia claimed that scientists are close to being able to accurately determine a whale's age from DNA analysis of a skin flake sample.

The strongest indication that this was not about research, however, came from the fact that at every IWC meeting the Japanese lobbied to lift the moratorium on commercial whaling. JARPA was set up on the heels of the moratorium in 1987, with the job of keeping the whaling fleet afloat and the market for the meat simmering along until the moratorium could be lifted. In the past eighteen years Japan had given $160 million in fisheries aid to half a dozen Caribbean nations alone, with the understanding that these countries would participate in the IWC and vote with Japan to legalize commercial whaling. Tens of million dollars more went to countries in Africa and the South Pacific. Mongolia had even joined recently, at the behest of Japan. The whaling nation of landlocked Mongolia.

■ ■ ■

The great whales underwent roughly three periods of hunting.

The first, stretching back thousands of years, involved going after coastal whales in small boats and either herding them into bays for slaughter or spearing them with handheld harpoons. The targets of the herding method were the smaller whales such as the pilot, beluga, and narwhal. Hunters with handheld harpoons went after the slower, more docile whales such as the right and bowhead. Cultures around the world, from Japan to Greenland to the Caribbean, practiced primitive coastal whaling.

As these species were wiped out along the coasts, ships went farther afield and began the second broad era of deepwater, or pelagic, whaling. As early as the fourteenth century, Basque whalers in search of whale oil were hunting off the Grand Banks; and in the sixteenth and seventeenth centuries, Norwegian, Icelandic, Dutch, and English whalers were roving across the Atlantic. By the beginning of the eighteenth century, Americans joined the hunt, and the whaling fleets of many nations were pushing into the North Atlantic and the arctic, systematically devastating the populations of right and bowhead whales. The Pacific arctic followed. In 1700, an unfortunate sperm whale was stranded at Nantucket and rendered. The high-quality oil ignited the American sperm whale fishery, and led to the rise of the Yankee whaling fleet—which, at its zenith in 1846, comprised 730 ships that hunted primarily in the Pacific. Sperm whales supplied lamp oil; ambergris for perfumes; and spermaceti, a very fine oil used in lubricants and high-quality candles. The discovery of petroleum in 1859 drove the price of whale oil down and may have saved the sperm whales from extinction. (Today they are listed as endangered.)

The third period of whaling, beginning in the late nineteenth century and continuing today, involved a combination of engine-powered boats and harpoon guns with explosive tips. These allowed whalers to go after the faster baleen whales such as fins, blues, minkes, grays, and humpbacks. They were hunted not only for meat but for, among other things, oil for margarine and special lubrication products, and glycerin for nitroglycerin. Many of these species were driven

to the edge of extinction. By the late 1970s, according to some estimates, the humpback population numbered less than 5,000. The
western Pacific gray whale was devastated, and numbers now only
about 100 individuals.

The rest of the great whales are recovering slowly. Experts in whale
research and marine biology such as Roger Payne and Sylvia Earle
believe the great whales should never again be hunted, period. The
Japanese ICR claims that the smaller minke whale has proliferated
and filled in the niches of the larger, now rarer species, and can be
hunted sustainably.

Earle put it to me eloquently: "As supposedly intelligent creatures,
doesn't it seem odd that humans might think that the best way to
engage whales is to eat them? When our numbers were small and
whales were numerous, killing a few whales for sustenance for people
who had few choices about what to eat was a matter of survival. Today,
it is a matter of choice. Can commercial killing of whales be justified?
Biologically, ecologically, economically, logically, morally, ethically,
realistically it cannot—not now, not in fifty years, not ever. No matter
how you crunch the numbers, dining on carnivorous animals that
may be older than your parents is not commercially sustainable.
Whales are long-lived, slow-growing wildlife, unlike domesticated
animals that convert sunlight via plants to protein in a year or less.
Even the little minkes, numerous though they appear to be, are vulnerable to predation by humans. Like other wild creatures, their populations can accommodate storms, disease, natural predation, changes
in food supply and other pressures they have endured for millions of
years, but nothing in their makeup has prepared them for the new
perils we impose: depletion of their food sources, pollution of waters
they live in, disruption of their social structure, and most obviously,
killing them outright. It defies logic to think that mobilizing large
ships consuming large amounts of fuel with large crews traveling
large distances to bring whale meat back over thousands of miles of
ocean to satisfy the tastes of a small number of consumers that live
half a world away qualifies as a reasonable use of resources, let alone as
a 'sustainable' enterprise."

When I asked biologist and famous whale researcher Dr. Roger Payne if he thought whaling could ever be sustainable, he said, bluntly: "It would be the triumph of hope over experience to believe that if we let Japan, Norway, or Iceland start whaling that they will now abandon their seamless history of spinning, slanting, and lying about what they do, in order to get around what the vast majority of the world's people wants. It will just be business as usual. It is madness to believe otherwise.

. . . I believe that people are educable, and that a society which does not kill the largest, most complex animals around it for the most mundane purposes is likely to have a more luminous future than one for which all of animals are but fuel for its meat grinders.

I believe that if you were betting on which animals' brains were most sophisticated and therefore which animals had the most to teach us and the most to contribute to the joy of our lives, that you might select whales, elephants, apes, and wolves. I think that the killing of whales for food is no different than the killing of apes, elephants, or wolves for food. Considering how much complexity of structure and behavior these four species groups have, and therefore how much we could learn from them about living, that to kill and eat them is not much different from using the works of Shakespeare to light your fire. The sonnets make good kindling and lots of people have probably used them for such, but such people, I suspect, haven't left much of a mark on history."

After breakfast the *Farley* turned north again to get around the floating ice shelf the size of Rhode Island that we had almost circumnavigated once before. The ship was pitching hard, running with the south swells, which were steepening. A few miles off to port was the great ice wall. The leaden sky was moving fast and lowering down over the naked cliffs and the high flat-topped islands. The weather was changing.

By late afternoon we had come up and around the great cliff island, running west and pitching in the groundswell that rolled out of the east-southeast. And then we cleared the far side and turned again almost due southwest, going as fast as we could, making almost ten knots. The sun struggled to break free all day. Once it did, and when it lit on the worn planking of the decks, bleached as bone, and glistened off the patches of new black paint, it warmed more than the air. The first sun in over a week. The thermometer read thirty-two degrees, but the day felt much warmer. Deck hands stopped and closed their eyes, letting their bodies sway with the roll, and letting the sun touch bare necks and faces. The Monument Valley icebergs scattered across the horizon caught the sun too and blazed with a blue life that was unearthly in its intensity. We passed one closely, and there were seven Adélie penguins on its skirt, diving into the water and clambering back up like so many sunbathers trying to keep cool.

We were 130 miles or so from the target. Marc was harnessed up, over the side, kneeling on the little porch of his can opener and welding on more braces so that the thing might really tear a hole. Little Kristy leaned over the rail and whipped him with a wood-handled crop, for fun. Trevor burst more rusty pipes trying to feed the water cannons, and finally ran fire hoses to the guns. Colin and Steve worked furiously in the bosun's locker, building a downrigger system for their Zodiac, similar to the rig used in deep-sea fishing, where you trail a weight from which your line extends so that you can troll beneath the surface. They wanted to meet the *Nisshin Maru*'s propeller where it lives. They wanted to make it easy for the prop to grab the steel cable and strangle itself. "We're going fishing for whalers, right?" Colin said.

Joseph Conrad wrote in *Mirror of the Sea*:

Of all ships disabled at sea, a steamer who has lost her propeller at sea is the most helpless. And if she drifts into an unpopulated part of the ocean she may soon become overdue. The menace of the "overdue" and the finality of "missing" come very quickly to steamers whose life, fed on coals and breathing the black breath of smoke into the air, goes on in disregard of wind and wave.

A ship as heavy as the *Nisshin Maru*, with no headway, would founder badly in the violent waves of a gale. Conrad went on:

There is no mistaking that sensation, so dismal, so tormenting, and so subtle, so full of unhappiness and unrest. I could imagine no worse eternal punishment for evil seamen who die unrepentant upon the earthly sea than that their souls should be condemned to man the ghosts of disabled ships, drifting forever across a ghostly and tempestuous ocean.

Watson and Alex were in the radio room reviewing the posting Greenpeace had put up on its website the day before. It included an instructive video that showed the *Esperanza* positioned to block the slipway. The slipway was the large ramp cut into the stern of the *Nisshin* up which the whales were hauled. Orange Zodiacs ran up under the looming black sides of the *Nisshin*'s hull and got thrashed and filled by the powerful water cannons. One of the Zodiacs had a banner that read, "Defend the Whales." The video showed the catcher ship *Kyo Maru* running up alongside the *Esperanza* and brushing it, and the *Esperanza* backing off. There were some aerial shots of the killer boat's aft deck, whales stacked like logs five across, and of the *Nisshin Maru*—dead minke whales on her deck and blood running from the scuppers and pumping in a crimson gush out of the through-holes above the waterline.

Watson said of the string of ships, "Looks like a Christmas parade."

"Their ship is bigger than the whaler," Alex said. "Look at that. Greenpeace had a fucking chance and they left it."

"It's just a media hoax," Alex continued, disgusted. "Do you think they're leaving the area now that they have the footage?"

"That's where they really could have done some damage," Watson said, pointing to the mild bump of the two ships. "The can opener would've done some damage."

"It's the David and Goliath thing," Alex said. "They wanna look like victims. They've got to try hard to keep the Zodiac right under the hose, to fill it up. They could pull out of the spray at any time."

The video showed the Zodiac filling, and crew members on the *Nisshin* leaning over the high rails and trying to spear the inflatable's pontoons with long-poled flensing knives. It looked dangerous.

"I got wet defending the whales. We have to make a T-shirt that says that."

"This is going to work out well," Watson said. "Greenpeace gets beaten up by the Japanese, the Sea Shepherd comes in and beats up the whalers. Protects the little kid from the big bully." More blood running from the scuppers. "Do you see anyplace to drop butyric acid?"

"Right on the deck," Alex said.

"Yeah, but without hitting anyone?"

"Best thing is aim for them, then you won't hit them." I had underestimated the parched dryness of the Dutch first officer's humor.

Now the video focused on a big sign on the mother ship's deck that read, "Science Based Lethal Research." And now the giant block letters on the side of the ship that spell RESEARCH. It was Orwellian.

Watson looked at the signs and said, "OK, that's what they're doing. I guess we were wrong. I guess we can go home."

We were running southwest. The seas were on the port quarter, dark walls of water patched with ice that barreled in toward the high stern. The *Farley* came up out of the troughs like a spirited animal scrabbling over a ditch bank and nosed down the other side. The steepening waves slowed her to under nine knots. The clouds moved in again swiftly, but off to the south a ruddy glow like a brush fire lay in a phalanx of flame over the water. Its reflection spread toward us like a burning slick from some violent conflagration toward the pole.

Gunter stood at the bridge windows with binoculars and called out the deadly growlers to Alex, who steered with the yellow plastic manual control. At 1736 Watson came onto the bridge and said, "Our Greenpeace friend just called me. They're twenty-five miles south of where they were yesterday. 65°55', 146°33'." He plotted the new course.

Trevor breezed onto the bridge and said he wanted to find out the height of the bow of a killer ship in relation to our bow. He flipped through the laminated photos of each ship of the whaling fleet that Watson had in a binder. Aultman said he had found schematics of the whaling ships on the Internet.

"Good," Trevor said abruptly. "Find out the height. That's your job." He left the bridge.

Aultman left for half an hour, returned, and said he'd calculated forty-four feet from the top of the harpoon to the waterline, using the scale of the harpoon gun at three feet high. Watson played a loud sappy folk song about Sea Shepherd sung by a supporter from Washington named Dana Lyons. Out on the freezing forecastle Trevor, in only coveralls, was lassoing a black cleat with a loop of steel cable. There was a light snow blowing around and what looked like fog banks ahead. Every third wave, the bow plunged and sent spray across his target. That's why he wanted the killer ship's specs—he had seen how close the *Kyo Maru* had run against the *Esperanza* and he wanted his shot at lassoing the harpoon gun on the bow. Little Kristy was out there in the spray and the roll, cheering him on. Three fin whales surfaced right below them—two adults and a juvenile. The mist of their blows flattened northward on the stiff wind. When they dived, they showed the lovely notched flukes; and a flight of snowy petrels, maybe fifty, took off from the boisterous water and fell behind the ship.

Watson anticipated we'd be near the target in about twelve hours.

A few minutes later Trevor came onto the bridge.

Watson said, "Can you throw it?"

"It works, kind of. It's going to be a tricky throw. If you're there, run to the other side and get low."

Lincoln said, "You don't know how it's going to snap back if it breaks."

Trevor wiped the melted snow off his face with his forearm. "It can only do one of a thousand things."

"When I was in the merchant marine we had a guy get decapitated with one of those, a spring line. Keep it simple, just throw it." Watson looked up at the GPS. "OK, just about eleven hours now. We might end up getting home here about the beginning of the year."

On the way down to dinner I noticed Inde on a short stool in the bosun's locker teaching Mandy how to splice and eye the steel cable. They were making more prop foulers. On the corkboard was tacked a new *Scuttlebutt*, Watson's newsletter for the ship, headlined "*Tora! Tora! Tora!*"—the battle cry of the Japanese pilots who attacked Pearl Harbor. "The *Farley Mowat* will ram the whaler with the objective of damaging the harpoon platform. . . . We need to stop one of the harpoon vessels dead in the water. If we do, the other boats will be forced to come to the assistance of the damaged vessel. . . ."

Dinner was clamorous. The crew scoured their mugs of lentil stew with pieces of homemade bread, and chewed the potatoes in gravy with a certain gravity: expectation, excitement, maybe a little fear. Outside the portholes the sky was darkening, and the blooms of white spray off the waves were more frequent. The *Farley* lunged into the troughs.

Kalifi called an all hands meeting in the green room. Alex would address the crew. They packed into the bench seats, sat cross-legged on the deck, stood in the doorway to the mess. The only one who didn't attend these meetings was Watson, who kept aloof.

"We are about twelve to fifteen hours from the target," Alex said. The laughter died fast.

"Tomorrow morning at nine we're going to be there. Greenpeace just pulled away. They chose to do nothing. Obviously, we're going to have a different tack."

Silence.

"When we get there we'll put three Zodiacs in the water and a helicopter in the air. The attack Zodiacs are gonna be Wessel with Luke, Jeff, and Gedden. Colin with Steve, Joel, Pawel. The third will have media."

He gave orders to use smoke bombs, water cannons, butyric acid. "Try not to hit anybody with the acid, we don't want to hurt anybody," he said.

Alex looked around the cabin. He looked as if he hadn't slept in days. His painter's pants and black sweatshirt hung loosely on his frame. His buzz cut emphasized the shape of his pale skull. He didn't even notice that he was swaying in a small rooted circle in perfect counterpoise to the pitch and roll of the *Farley*. But his focus outward was fierce. I thought he and big round Watson fit together like a bow and arrow.

"If you use a prop fouler," he said, "don't get your hand caught." He looked at Chris Aultman on a bench seat at the near table. "We're gonna send you off tomorrow around seven a.m. Maybe six. Don't use the radio."

"If we are in a confrontation, then we can use the radio?"

Alex nodded.

Alex said, "There's gonna be teams. Watch each other. If you're in a Zodiac you've gotta pay attention; don't let people stay in the water."

Harpoon gun lassos, stink bombs, and can openers. The bludgeon bow of the *Farley* and the very real chance of collision on the high seas.

Jeff said, "Is the chance of action pretty high?"

"Yes, if they haven't moved. Helicopter's gonna be launched at six. Get the Zodiacs ready at six." He swayed with the ship and tapped the air with a finger. "Be *ready* at six a.m.; don't have breakfast at six. Be ready."

Now Pawel. He was such an odd fish. Impish smile, rapid blinking: "If we have the opportunity to snatch a sample of a Japanese scientist, should we?"

Alex said, "I think if it's nonlethal research it'd be OK," he joked. Then he clarified: "No, I don't think Paul wants us to board any ships."

Allison said, "The paint locker. That will be our brig." She was serious.

Mandy, always serious, said, "What is the policy if the Japanese board us?"

"Kick them the shit off," Alex said. "That goes for Greenpeace too. If we end up with one of their Zodiacs, they have to give us an apology for missing an opportunity yesterday."

Big Chris Price, the .50-caliber man, said, "If the Japs try to board our ship, everything short of killing them, that's all right."

"They shouldn't be expecting manicures," Jeff said seriously.

"OK," Alex said. "Try to get some sleep."

That was it. As soon as the meeting broke up, the frenetic clamor of soldiers preparing for battle took over the ship. All night the smell of burning welding rod, the hammering, and grinding filled the hallways. We had come 2,000 miles and the target was hours away. Pawel approached me with a pair of split fin flippers, a snorkel, and a mask, and asked if I would be the rescue swimmer if someone went overboard and was in trouble. I agreed and followed Alex up to the bridge level. Watson wanted to meet with his officers in the chart room to discuss the attack. Through the doorway to the bridge and out the windows I could see the fog closing in again and the bow plunging in spray as the sea got rougher. I could see the potentially deadly chunks of ice on the walls of the swells.

Watson said to Hammarstedt and Alex, "We can actually destroy the whole front end of our ship without sinking it."

I glanced again at the weather and the seas and imagined the *Farley* plowing into them with a bashed front end. Not pretty. But Watson was serious now. He would do it. He had said repeatedly that the ship was expendable, and he had played chicken enough, and used his ships as battering rams enough, to prove the point.

Watson said, "It seems to me if we can ram into the slipway—call it Operation Asshole—we ram it at full speed, come down on that point off a wave, then they have a structural hole. The slipway comes down right to the water; if we come down right there it will compromise the integrity of the hull."

He had talked earlier about the danger of the can opener to the other ship, saying that above the waterline it posed no danger of sinking the attackee. He had said he would never threaten a ship with sinking on the high seas. But now he was talking about cracking the *Nisshin*'s hull right at waterline. A "structural hole" at waterline is what a torpedo tried to do. He wanted the ship out of action at all costs, and it seemed that cracking the slipway would do the trick.

Watson leaned against the chart table. He was looking dapper in wide-wale corduroys and a fleece sweater, and I noticed that his hair was freshly washed and brushed, an anomaly one noticed on the captain. His captain's quarters had a full bath, but Allison complained that she could rarely get him to stop long enough to bathe. Apparently he viewed the coming showdown as a special occasion. He motioned with his beefy hands, "The harpoon and slipway, they are the two targets." That meant two different ships, as only the killer boats had harpoon guns. "If they can't bring a whale up the slipway, that's the end of their whaling. Ram that thing so far up the slipway we stick."

The new intercept point was forty-three miles away.

"We'll launch the chopper in four hours," he said.

14

Force 7

December 23, 2005, 0245

The ship was pitching more heavily. The wind had swung around to the southeast and increased in speed, and the swells were bigger, maybe fifteen feet. Aultman's crew was out on the heli deck struggling to strip the chopper of the blade covers, checking fluid levels. They were harnessed up and tethered to taut lines and the long red covers were whipping wildly in the wind. They held onto each other hard when someone lost balance. The ship labored directly downwind at not much more than an idle, and the big swells rolled under the stern and pitched her head down, surfing the face, then shoved her nose to the sky and she floundered in the trough. The tops of the waves erupted in white, and spumes like spittle threaded the swells downwind. We were two miles south of the intercept point. There was nothing on the radar, and the snow was blowing sideways.

At 0303 Hammarstedt came onto the bridge where Watson was settled in the captain's chair. He had his orange-and-black U.S. Coast Guard Mustang suit on, and his feet were propped on the sill.

"We're not going to stop, Paul?"

"No." Watson glanced at the radar.

"We don't know where they're headed. They might be going southeast. The helicopter crew is getting ready to launch the chopper."

"OK."

Aultman squeezed through the hatch into the chart room a minute later. He fought the wind to open the door and it slammed shut behind him. He went straight to the chart. Watson met him there.

"OK," Aultman said. "So I go twenty miles west, then thirty-nine miles south, then back along your course of 210 degrees until I find you."

"Right." Watson would go back down the fleet's projected course, hoping to run into them head-on. Aultman would hit that line to the south and run back up along it, scouting. Our new course, SSW, would put the weather almost square on the beam, where it would roll the *Farley* without mercy.

"OK," Aultman said. "If I find them, I mark it on the GPS, get my ass back."

"Right. If there's an emergency and Greenpeace is there, I'm sure you can land on their pad."

Aultman took off his glasses and wiped them on the collar of the sweater under his Mustang suit. "OK, you're two miles off, east of that line. You're the only black ship on the ocean. I hope I can find it." He turned to George, the videographer who was going with him this time. "All we have to do is get that door back on, start the start-up sequence, and we're ready to go."

Bright fog rolled over the water. The snow squalls moved fast, banks of darker cloud that came right down to the sea. When they overcame the *Farley*, the snow thickened and flew sideways across the decks. The flakes were very cold and dry.

Aultman zipped the top of his suit to his neck, tucked the GPS into a pocket on his thigh, and went back out the door. Through the shuddering of the gusts against the superstructure I heard the engine of the chopper start up. On the bridge, Hammarstedt was peering intently at the main radar on the port side of the console.

"Every once in a while we switch it to scan out to forty-eight miles, but it's pretty much impossible to see beyond thirty miles."

We were about 240 miles due east of the south magnetic pole and about 120 miles west of the Mertz Glacier. Aultman came back onto the bridge.

"Hey, can we turn 180 degrees? Because we're going with the wind right now." He went back out. Hammarstedt lifted the manual control and brought the *Farley* around. Halfway through the turn the waves came onto the port side. The *Farley* rolled hard, maybe thirty degrees, and I tumbled across the chart room, along with two plastic coffee mugs and somebody's life jacket. The ship came up into the wind and waves and steadied. Now we were heading 140 degrees southeast. The snow drove into the windows of the bridge and the whitecaps churned patches of aerated green out of the dark waves.

At 0345 Aultman came onto the bridge and said, "Paul, we're not going. It's too rough. About a forty-five-knot wind."

"OK." He said it with no emotion. "Peter, resume the course of 210, back down the line."

Hammarstedt swung the *Farley* back to SSW, and as she came down off a steep roller and yawed to port she took the next wave heavily and slammed over. Then she settled into a hard roll as the weather came on to our port side. The back sides of the waves that rolled under us were strung with foam. The snow lightened and blew across the main deck and canting forecastle, thickening as we passed through blowing clouds. Hammarstedt went off watch at 0400 and Alex came on with his crew. Watson went down to his cabin to catch a few more hours of sleep before he came on again at eight.

Down in the mess, Watson had written on the eraser board: "*No desit virtus*—let valor not fail."

At 0800, refreshed, Watson took his watch. He was still in his Mustang suit, which meant that he still held a reasonable expectation of action. The suit was bulky and hot, and you didn't wear it unless you thought there was a chance you might end up in the water.

"I guess if it was easy, everybody would be doing it," he said, lumbering into his chair. "Gilda Radner once said, 'It's always something. If it isn't something, it's something else.'"

His stony detachment had been transmuted into his usual upbeat playfulness. Allison had said that he had no emotional range, that it was a form of protection. Casson had added, "He goes from grumpy to surly and back." Now that I was used to his moods, I thought I had seen him go from fury to euphoria.

Watson handed me a broadside he had just sent to a core group of Sea Shepherd's supporters.

Sea Shepherd Prepares to Attack the Japanese Whaling fleet in the Antarctic Whale Sanctuary: Message from Captain Paul Watson on board the Sea Shepherd ship *Farley Mowat*
December 23, 2005

I am sending this e-mail from our ship the *Farley Mowat* to a small list of friends and supporters.

We are down off the coast of Antarctica about 180 miles off the Mertz Glacier and the Adélie coast.

We are about five hours from interception of the Japanese whaling fleet. We are presently on an interception course. . . .

I want you all to know that I'm down here with forty-three dedicated and courageous volunteers who have given up their holidays with friends and family to be here to defend the piked and fin whales from the merciless grenade-tipped harpoons of the Japanese fleet.

We anticipate a battle today. I've been working toward this showdown with the Japanese fleet for twenty-five years. Now at last, their ships are within striking distance and we will do everything we can with the resources at our disposal to shut down their illegal slaughter of these gentle and intelligent creatures.

We may lose our ship and find ourselves in our lifeboats within the next few hours. I'm quite sure we will sustain damages.

But I want you to know that there is nowhere in the world that we would rather be at this moment, and there's nothing else that we can imagine doing other than what we're doing right now.

For this holiday season, we want to give a gift of life to the whales and if we can stop this fleet, if we can stop the killing, we will be very happy. . . .

I hope that within twenty-five hours, if we still have a ship and if we still have communications, I will be able to report to you the consequences of our intervention.

Happy holidays,

Captain Paul Watson and the crew of the Sea Shepherd ship *Farley Mowat*.

Aultman came on, bracing against the heavy roll. He looked out the buffeted windows at the prospect of the building storm. He had gotten some sleep and he looked relieved. "It was probably the right decision. The snow, coming in sideways, looked like little wads of Styrofoam. Forty- , forty-five-knot wind, snow flying everywhere. All of us trying to tie it down."

Watson went back to the radio room. The ship continued down the intercept line. Watson e-mailed his informant on the *Esperanza* that we were in place to intercept but that the fleet must have altered course, slowed down, or stopped, as we hadn't seen it. He pulled a newspaper story off the *Melbourne Age* website and announced that the *Kaiko Maru* would be arriving on Christmas in Hobart, Tasmania.

He read to us from the article:

. . . Sea Shepherd has a record of incapacitating vessels without injury. . . . Direct action against the whalers is also set to rise, with the hard-line group Sea Shepherd likely to arrive off Commonwealth Bay today, saying that unlike Greenpeace it intends to stop the fleet from whaling.

Prime Minister John Howard said that at a recent meeting

with the Japanese Prime Minister Junichiro Koizumi he had raised the whaling issue.

"I did not lose the opportunity of telling him of my continued opposition to Japan's position on whaling," he said. "However, I do not support action which endangers lives or breaks the law."

. . . Japan expects to take 935 minke whales [and] 10 fin whales this summer under a self-awarded scientific permit outside a global moratorium on commercial whaling.

Tension is expected to rise with the arrival of Sea Shepherd's ship the *Farley Mowat*. Its captain, Paul Watson, has a long history of incapacitating whaling ships, including ramming and sinking them, but without injury or loss of life.

Watson handed me the sheet. "If they were Filipino or Indonesians, Australia would arrest them as pirate whalers. It seems there's a double standard here. When it comes to Japan, we'd better not send the wrong message. In other words, we've got an ass to kiss and we better kiss it." Swinging back to rage.

"If we know they were there ten hours ago, they can't be more than 100 miles from here. My friend on the *Esperanza* said he'd call me this morning.

"If the Australian navy comes down to arrest us for committing a crime in Australian territory, that'd be great. 'Gee, well, we're going to arrest you because you recognize our territorial claim, but we're not going to arrest the Japanese because they don't recognize our territorial claims.' I just got an e-mail from Carl Pope, the Sierra Club executive director. He's off to Bethlehem for Christmas—"

"I just got a new target," Lincoln said. He was standing ramrod-straight, hands braced on either side of the main radar hood, looking down at the screen. Watson ignored him for the moment.

"Carl Pope told me, 'You've got to apologize. You made a derogatory statement about Christians.' I said, 'I did that on my own time—'"

Lincoln, still completely focused on the radar, said, "The target has dropped to two and a half knots."

The *Farley Mowat* on the hunt in Antarctica. The chopper is off for a long scout ahead. *(Photo © Paul Taggart)*

The *Farley Mowat*, docked in Melbourne, early December 2005.

Left to right, engineer and punk rocker Luke Westhead, and the J. Crew, semi-professional gamblers from Syracuse, Jeff Watkins, Joel Capolongo, and Justin Pellingra.

Bosun's Mate Colin Miller, *left*, and deck hand Joel Capolongo, preparing the ship for departure.

Two Dutchmen: welder Marc Oosterwal, *left*, and ship's artist Geert Vons.

Top left: Quartermaster Lamya Essemlali, from France.

Top right: Second Officer Peter Hammarstedt, the Swede.

Above: The Fourth Estate: videographers Ron Colby, *left*, and Kristian Olsen.

Right: Dutchmen Marc and Geert with a time capsule they built. Inside are letters and messages to future generations. The capsule was left the next year on a tiny rock islet in the Southern Ocean called Scott Island.

The Can Opener: chief engineer Trevor Van Der Gulik preparing a support for the weapon. He is being belayed by engineer Willie Houtman. Welder Marc looks on. The can opener is a seven-foot steel blade designed to tear open the hull of another ship.

The crew assembles on deck. *(Photo © Paul Taggart)*

Weapon making 101: deck hand Inde—aka Julie Farris—shows another hand, Mande Davis, how to splice an eye into a steel cable. They are making prop foulers, a weapon designed to disable another ship's propeller.

Melbourne local electrician and union boss Sparky Dave DeGraaff. Dave saw the *Farley Mowat* from the top of one of the new buildings under construction in the harbor and signed on.

J. Crew Jeff with the "Jellyfish," a prop fouler with multiple cables.

Captain Paul Watson with his wife, and galley assistant, Allison Lance Watson. Allison is a fierce animal rights activist who once leapt with a knife into a net-caged bay in Taiji, Japan, and freed dolphins penned there for slaughter.

Chief engineer Trevor Van Der Gulik. Highly trained, he has supervised teams of 500 ship's engineers in Dubai at the largest shipyard in the world. He is Watson's nephew.

The chopper lashed to the heli deck in big seas, Christmas Eve, 2005. The deck is about twenty feet off the water and the waves tower above it. *(Courtesy Chris Aultman)*

The *Farley* bears down on the *Nisshin Maru* in a fierce Christmas gale. *(Photo © Paul Taggart)*

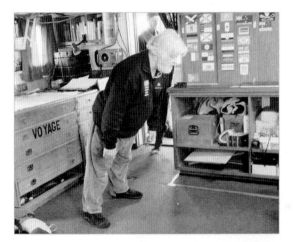

The captain loved competitions. Here in the chart room, he bests all comers in the no-hands game. He is actually standing perfectly vertical; the ship is taking a heavy roll in thirty-foot seas.

Bosun Kalifi Ferretti-Gallon untangling line confiscated from the Uruguayan longliner *La Paloma.*

The chopper coming in over the heli deck on a ship beset with ice.

Returning to the *Farley* after a three-hour scout along the ice edge with helicopter pilot Chris Aultman.

Ship's Nurse Kristy Whitefield, an emergency room nurse in Melbourne who joined the ship just before departure.

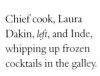

Quartermaster and Earthfirst! forest activist Gedden, aka Jon Batchelor.

Chief cook, Laura Dakin, *left*, and Inde, whipping up frozen cocktails in the galley.

Approaching an
island of ice.

Geert's back, after he went swimming
New Year's Day, 2006.

Each crew member who
swam got an official Penguin
Certificate from the captain.

Dead juvenile sperm whale.

The bow of the *Farley*. In the foreground are the three Zodiacs lashed to the main deck. To the left, the port davit, or crane. On the forecastle head is the windlass for bringing up the anchor, the ship's bell—taken off an Iraqi freighter and once rung by Saddam Hussein—and the water cannon at the bow.

Launching a Zodiac
for action.

South African Zodiac
pilot extraordinaire
Wessel Jacobsz.

Trevor snowboarding
Antarctica.

Two minutes later four of the crew were in the water, testing wetsuits donated by the surfers of Australia.

Quartermaster and news reporter Emily Hunter. She carried her father's ashes to Antarctica and spread them on top of an iceberg.

Captain Watson after dinner in the mess. To the right are Flying Inflatable Boat pilot Chris Price, and Inde. Behind is quartermaster Darren Collis.

Another monument of ice.

Watson in his captain's chair and the U.S. Coast Guard Mustang suit he bought on eBay.

Joel after hours of harrying the *Nisshin Maru*. Wessel took his Zodiac up under the bow of the factory ship where his crew threw prop foulers. *(Photo © Paul Taggart)*

Giving chase to the *Oriental Bluebird* after the ramming. She fled north, never to rendezvous with the *Nisshin Maru* again. The "Whale Meat" on her stern was painted by Greenpeace activists during a transfer of whale meat from the factory ship to the *Bluebird*.

The *Farley Mowat* patrols the seas alone. *(Photo © Paul Taggart)*

Watson pointed ahead. "There's a whale!"

Half a dozen minke whales were running across our bow. Running, as fast as a minke whale can go. Each one breached partway, getting its head out for a fast look at us. They were running from west to east. They blew and breathed about every five seconds, a constant mist of staggered plumes torn away by wind. They looked terrorized, panicked. Their fear hit the bridge like a blow. It must have been like the instantaneous transmission of emotion Watson had described in his encounter with the big gray whale as it died.

Watson said, "I've never seen them move like that before—they're running scared. They seem very agitated, moving fast, blowing every few seconds."

We watched them move off to port, into the waves and fog. All of us on the bridge were thinking the same thought, that they had been attacked by the whalers and were fleeing. The catcher or killer boats could go twenty-five knots. A minke whale can swim in bursts of twelve knots. They could run as fast as they were able, but if the *Kyo Maru* or *Yushin Maru* saw them and wanted them, there was nothing they could do to stop it.

The groundswell was black walls of water now, looming to the level of the bridge deck about twenty feet high. The wind took off the tops of the waves and hurtled the spray against the superstructure. The little thermometer screwed to the bulkhead still showed thirty-two degrees, but the driven wet felt much colder. We looked to be sailing directly into a darker cloud front of snow and fog. Whatever this was, it was not abating, but building.

So was the sense of frustration, of fury, on the ship. We had come within hours of the whaling fleet and missed it. Because of wind and snow. We were casting back and forth like a hound that has lost the scent. It was two days before Christmas, and even vegans miss their families. I know I missed mine. It took a force of will to convince myself we were not on a fool's errand. I thought how completely vulnerable we were to everything: to the caprices of the weather, to our old and fragile equipment, to the inexperience and idealism of the crew who would jump into a Zodiac and overload it at a critical moment in the chase, and to the speed and efficiency of the Japanese fleet. It seemed that any of 100 things could imperil the mission or our lives.

Another group of minkes, four this time, ran close past the ship, blowing without letup, also heading east. A black-browed albatross came up out of a trough and caught the gale and rocketed up and over the bow.

Just then Lincoln said, "Got a target! Radar's giving me seventeen and a half knots."

"That's the speed of a whaling boat," Watson said.

"Two hundred eleven degrees heading away from us. Seventeen nautical miles, speed now fifteen knots."

Watson waited.

"Now they're doing twelve, now eight, now five, three, two."

"Must've got a whale."

"Twenty-one miles away. Now six, now back up to thirteen."

Lincoln: "Three, six, five, six, four, three, two, 2.3, 1.1."

Now the target was just sitting there, going half a knot, looking more and more like an iceberg as we got closer. Again, the lift of adrenaline and the letdown of a sea that was empty of whalers.

Watson wasn't ruffled. "Listen—" He read from another news report he had pulled off the web earlier:

> The Federal Minister for the Environment, Senator Ian Campbell, should put the Japanese whale-killers out of action when the *Kaiko Maru* docks in Hobart tomorrow, Greens Leader Bob Brown said today.
>
> "The ship coming to Hobart is part of a fleet harpooning our whales in our territory off Antarctica. The government should impound the ship as it does Indonesian fishing boats and the pirate Portuguese Patagonian tooth-fish ships." Watson continued reading. Senator Brown went on to say that the Japanese ship was expected to dock at the Self Point Oil Wharf at 8:00 a.m. Saturday, December 24, and should not be allowed to return to the kill.

Watson quoted: "'The government has dedicated millions to tracking down fishing pirates so it has no excuse for hosting whaling pirates,' Senator Brown said."

Watson grunted his assent. He was pleased that at least some Australian government officials were showing some spine.

We chugged on toward our ship of ice. The seas were approaching twenty feet. Can a ship be glad? Maybe not. The *Farley Mowat* met the side-on waves with a deep-bellied buoyancy. Most of the time. Sometimes she would yaw just wrong coming down into a trough and the next roller would slam her to starboard and you could hear sundry objects hitting the bulkheads and cabinets and the roll gauge would read thirty degrees or more. Kalifi went onto the forecastle dressed in light cotton pants and with no foul weather gear, unharnessed, and was trying to fit a fire hose into the shooting end of the center water cannon; I guessed to back-test the pipes. She could barely stand in the wind. Her hair flew and she almost got knocked off her feet in a gust. She clutched at the cannon with one white-knuckled hand, then gave up.

Alex and Hammarstedt came on to the bridge, the first mate with a mug of coffee. He said to the second officer, "If you see Kristy, Dave, Mathieu, I need their emergency contact info."

Watson glanced at his officers. He said to Lincoln, "Maybe when we get up to that iceberg we can get around it and get the helicopter up."

"Yeah, six and a half miles to the iceberg."

"Fourteen miles to where they were last seen," Watson said.

"Visibility is better," Lincoln said. "Wind still a bear. Kalifi was definitely struggling to stay upright."

Watson craned around to Hammarstedt. "What are you doing up?"

"You know how it is on this ship—I heard we had a target on the radar, whales in distress."

Now Trevor made his way up on the forecastle. He went from rail to stanchion to guy wire, like a herky-jerky square dancer being passed from hand to hand in a lunatic reel. When he got to the cannon he swiveled it, aimed it over the starboard side, and let loose. A flaccid stream arced over the side and wavered in the gusting wind.

"That'll show them," Watson said. "We piss in your general direction."

The iceberg Lincoln had mistaken for a whaler was a flat-top, a mile long. The floundering *Farley* went for it the way a skier beelines for a mountain hut in a storm. We passed two Adélie penguins on a floe the size of a Zodiac. They stood side by side like figures on a wedding cake. Watson told Mia at the helm to go around it on the port side and tuck in behind. Allison hugged her husband in his U.S. Coast Guard exposure suit and said, "We're ready for battle, huh, baby? Bring them on."

The *Farley* swung up into the lee of the big iceberg, and the bridge went quiet. The wind that had been battering the superstructure was gone. The creaking and whistling. The heavy rolls. The *Farley* nosed up into water that had gentled to near smoothness.

Alex stepped onto the bridge. Watson said, "They can't be more than fifty miles from here. We're seven miles from where they were yesterday."

Aultman came onto the bridge shrugging into the arms of his Mustang suit.

"What's it look like, Chris Aultman?" Allison said.

"Looks great; let's do it." Did he know what the wind would be like as soon as he cleared the top of the ice cliff? Watson asked him to make a sixty-mile-radius arc west and south.

"We're about 180 miles from the French station," Watson said. In case Aultman got into trouble.

"OK." Aultman went aft to strip the chopper of its covers and get it started. His deck crew, scattered over the ship, put on foul weather gear and gloves and headed for the heli deck. Casson, who had been in the galley, stopped at the dry-erase message board on the way out. Below Watson's line in Latin, he wrote, "*Nil illegitimus carborundum*— Don't let the bastards get you down."

The chopper wouldn't start. The battery seemed dead. It needed a jump. Quiet George was scheduled to fly and film today, and I was kind of glad it wasn't me.

Alex stood beside his countryman, Geert, who was drawing anti-Japanese T-shirt designs at the chart table. This one was a yin-yang with half the circle an arched sperm whale harpooned and spouting blood.

Alex said, "The battery was not charged."

Geert rolled his eyes sideways while he filled in the yin with red pencil. "Wasn't that a bit stupid?"

"I don't make comments about your drawings."

At 1230 Dennis blew a gust of cold air into the chart room and reported that the battery was fine, it just wasn't getting any juice to the motor. Half an hour later, George said, "All fixed. Just a wire that had eroded away."

Oh, just a wire. Eroded away. I watched George closely. He didn't seem to be muttering prayers under his breath. Aultman was beginning the start-up sequence, so George zipped into his dry suit, picked up his camera, and went aft. They took off at 1330. Aultman cleared the berg and steadied the helicopter in the blast, then rose and accelerated downwind like a petrel. Even the two penguins lifted their beaks to watch. They flapped their wings madly.

If the Japanese whaling fleet was over the horizon, just beyond the reach of the radar, tracking northward, then we would know in less than an hour. Aultman, eschewing the marine radio so as to keep our presence secret, would call Watson on the Iridium sat phone. The fog had been torn away by the wind, and though there were fast-moving dark snow squalls that came down to the water, the visibility was not bad.

The crew became jocular when they had to wait. It was Hammarstedt's watch, and the three Dutchmen joined him on the bridge and began to torment him for being Swedish, recapitulating some ancient lowland enmity which I suspected had something to do with Scandinavian bloodlines being forcibly introduced into their own at some point in the horn-helmeted past.

"If you visualize your language," Geert said to the second mate, "it is the color pink."

"At least it is not a throat disease," countered Hammarstedt, at the helm, holding the *Farley* up under the berg in its sheltered berth.

"Yah, but I like the Swedish women," crowed Marc.

"Because they are tired of the Swedish man," said Alex.

"Yah, drinking all the time. The Vikings. Destroy all villages, deflower virgins. Bad people."

Hammarstedt said with his cryptic smile, "Unlike you, we didn't have a single slave; all we did was slaughter people."

Watson was in the radio room, keeping up his end of the media war. This time the target of his scorn was Ian Campbell, Australia's beleaguered minister of the environment. Campbell had been consistently against commercial whaling, as had his government, which for decades had been part of the core antiwhaling group in the IWC, led by the United States, England, and Australia. But that wasn't enough for Watson, who knew that Australia was one navy frigate ship away from shutting down the entire Japanese operation in the Australian Antarctic Territory.

Watson's fingers flew over the keys as he composed a press release:

> If his authority exists because we are in Australian Antarctic Territory, then why is Australia doing absolutely nothing about the invasion and exploitation of this territory by the Japanese? Ian Campbell should concentrate on being the minister of the environment and protecting whales instead of acting like a cheap public relations flack for the Japanese whaling industry.

This was more than criticism: Watson was taunting Campbell, egging him on to send a navy ship to intervene against Sea Shepherd. That would be a public relations coup Watson could milk for months.

A force 7 gale was raging on the other side of this cliff of ice, two little Adélies were squawking at each other in envy of our helicopter, there was a big blade welded to the bow and the ship was hiding in the radar shadow of the berg, waiting to pounce on a whaling ship. Meanwhile, Watson was in the cramped radio room firing off one media salvo after another—to Greenpeace; to his website; to newspapers in Australia, New Zealand, England, and the United States.

He came out of the radio room shaking his head. "Did you see the new stuff on the Greenpeace website? Some Japanese guy hit one of them over the head with a pike yesterday—with the end of one of their own banners." He went to the bridge windows and looked out at the storm beyond our wind shadow. "Still no coordinates, but they do say they're not far from Commonwealth Bay, which is 120 miles from here. So they're around. They are trying to root out the leak on their boat." For some reason the Dutchmen thought that was hilarious and laughed heartily. Maybe it was the word "leak"—they came from a country of dikes.

"They haven't figured out it's him yet. Keep your fingers crossed."

At 1420 the leak e-mailed Watson to say that the Japanese were now at 63° 44′ S, 146° 31′ E—120 miles due north of us.

"That means they would have passed us in the night over the horizon," Watson said. "Too bad we couldn't get the helicopter up this morning." If we had gotten the helicopter off at 0245, we would have seen them.

Hammarstedt went to the chart and, after a minute of figuring with ruler and pencil, said, "We passed within about twenty-two miles of each other at seven this morning."

In the storm, with the heavy seas and fog and all the ice around, the watch hadn't been able to see them on radar. We were chasing shadows and reflections.

Watson got on the sat phone and called the helicopter back. As was typical, he registered no disappointment. Rather, he turned to the ice cliff face of the berg, which was crazed with a diagonal blue fracture. "I bet if you put a shot right there, that big piece is going to come down." He was still in his Mustang suit, ready for action—unwilling, I guess, to let the moment pass without some sort of rough fun.

"You got your gun," I reminded him.

"Never fired it—I wonder if a parachute flare would bring that down." He disappeared into the radio room and we heard drawers and cabinets opening. He came back clutching a couple of signal flare tubes. "We got so many of these things," he muttered. He went onto the port bridge wing, aimed one, and fired. It went straight into the water with a sizzle. "I'll try again."

The berg was 100 yards away. He pulled the string. The flare arced up, maybe eighty feet, and floated down on its little parachute, a red phosphorus flame trailing white smoke.

"Hmph," Watson said. "Not very accurate. It'll be interesting if the helicopter saw that—might come in handy." He looked again at the cracked wall. "Wonder if the fifty-caliber would take that down." I wondered myself if all this was just a way of easing the big gun out of its hiding place without alarming the crew, and of getting some practice before any encounter with the whalers. Or was it just Watson's inability to resist fun?

"Go for it. It'd be cool. Let's go," I said.

"It's Chris Price's call, I've never fired it. Don't let Willie near it."

I thought it was missing a part and wouldn't fire.

Aultman took the chopper in over the stern just as a snow squall overtook our haven in the lee of the iceberg. He was pushing his edge. He hovered, unsteady in the new gusts, watching his timing, and then dropped the chopper onto the deck. As soon as the heli crew had the pontoons strapped down, Hammarstedt engaged the propeller and maneuvered the *Farley* back out into the waves. He placed three new numbered wooden blocks in the bearing slot beneath the central bridge window, showing the helmsman the new course. Now they read 359, almost due north. Three yellow rubber ducks had been placed just above them on the sill. They looked eagerly ahead. Where the hell did all the ducks come from? They were multiplying.

The first waves caught the *Farley* on the starboard stern quarter and rolled her hard. They were often up to the bridge deck now, over twenty feet, and sometimes a blackwater wall obscured the horizon as the *Farley* dropped into the trough. In this weather, with any course that wasn't directly into the sea, or running with the waves, everything was a challenge. The galley crew kneaded bread and chopped turnips while bracing themselves into a corner between counter and bulkhead. Dishes clattered in the drying boxes, and Allison attacked them with a gymnast's athleticism, her back to the mess, grabbing the counter rarely, the ridges of muscle on either side of her spine flexing through her blue undershirt. Walking down the long companionway, if you were in the zone, you could lean at a cartoon angle from side to side while the boat rotated around you and never miss a step. Even going to the head was an adventure. You hoped the seawater wouldn't slop out of the bowl, and you held tight to whatever you could. And sleep—forget it. Every muscle in your body went taut, shifting and pressing the raised sides of the bunk in an effort not to go airborne.

At 1510 Watson was in the mess having a cup of tea. He had pulled out the little poker case, hoping to catch some media suckers in a game of Texas Hold 'em. When he shuffled the cards the J. Crew glanced over and licked their chops. The captain cocked his head and motioned them over. "We're not allowed to play cards during the day, captain," Joel said with his best soft-spoken gunfighter's civility.

"Captain's exemption; get your money."

At 2130 the whale alarm sounded. Three humpbacks—two adults and a calf—swam across the bow, breached one after another, did a slow lap of the ship, and moved off. The calf didn't really breach, just lifted its head to take a look at us. I hoped that in 2007 when the Japanese had humpbacks on their kill permit, these whales would be more circumspect about approaching a boat like us—we were just about the size of the catcher ships. Or maybe they knew. I thought they probably did. Not that they were safe this season: a sampling of whale meat at the Tsukiji and other markets between 1993 and 1999 found blue, sei, humpback, fin, and the critically endangered Western Pacific gray whale, of which only 100 remain. With whale meat at ten or fifteen dollars a pound, I wasn't sure what kept the Japanese from taking any whale that crossed their path.

When I went to bed just before 2000 we were at south 64°50'.

Saturday, Christmas Eve, 0500

We had halved the distance to the fleet's last position since last night. Still about sixty to 100 miles to go, depending on how fast they were moving and if they held their course.

On the bridge, Alex, Gedden, and Gunter maintained a silent vigil. When the ship pitched to the top of a swell and we could see out to the horizon, the whitecapping ocean looked clear of big ice. Gunter kept an eye on the radar, which was set out to thirty miles.

Trevor came up after 0700, clean-shaven and fresh in clean blue coveralls. His blue eyes swept the gale-raked sea. He told Alex that he'd just sounded the tanks and we had thirty-one days of continuous motoring left. There were seventy-nine to eighty tons of diesel fuel in the seven tanks, and we used about two and a half tons per day.

"In thirty-one days we'd better be in port," Trevor said.

Watson came up for his watch at 0800.

"Our spy says the fleet is moving south, moving west, says it doesn't make any sense what they're doing. They are not whaling. They haven't whaled since they started running. He says his captain believes the *Nisshin Maru* won't hesitate to ram us if they have the chance. I think that would be interesting. See what the Australian government says then."

He checked our position and the chart, then told Lincoln to change course to 300 degrees. He wanted to come across, west, at a shallower angle, staying south of the fleet in case it should break back for the continent, head back for the ice where the whales were. So now we would head almost directly with the storm and the seas, which would make the ride a lot more comfortable.

Alex and Trevor retired to the chart table to study our options.

"Be nice to have an hourly update. That'd be perfect," Alex said.

"Yeah, two times a day. That's pretty good. I like the idea of working together. The officers are great. Most of the crew are like, whatever. They're all on that boat because they can't get on this one."

Alex smiled. I wondered how the Greenpeace crew really saw Sea Shepherd. They were getting seventeen dollars an hour to crew, so they were paid to be loyal. Undoubtedly they had the integrity of their own pacifist mission, because they took great risks pounding their Zodiacs up alongside the whalers who were flailing down at them with flensing knives on poles, and shooting harpoons close over their heads.

They studied the chart. Alex said he liked the idea of moving south and waiting for the fleet to come back to a good whaling ground. But it could come back to the coast at any longitude. Nobody knew what bay the Japanese would decide to hunt. And anyway, Watson was heading WNW. He had his own ideas.

Watson took his captain's chair and gave all of us on the bridge an update on the media war. He said there were new articles in the *Age* and the *Sydney Morning Herald*, which he flourished. The *Herald* story opened with this lead:

> The hard-line antiwhaling group Sea Shepherd yesterday raised the stakes in the Antarctic whaling crisis, saying it was about to attack the Japanese whaling fleet and could lose its own ship. . . .

Watson cracked his knuckles. "I never said we would attack them!" Incredulous. Then: "Of course we will attack them, but they don't know that! Oh, listen to this":

> Federal Environment Minister Ian Campbell said the statement was "quite scary." He said he had passed it on to Attorney-General Philip Ruddock and Justice Minister Chris Ellison for examination.

"There appears to be a prima facie case that they may be setting out to break the law," Senator Campbell said. "I think there is a very good distance between this and the generally positive approach by Greenpeace. I think what Greenpeace has been doing is a service to the cause," he said. "But if Captain Watson does what he says he's going to do, it will set the cause of whale conservation back for decades."

"Oh, this is better. Here's the Rat":

Greenpeace expedition leader Shane Rattenbury said, from the Southern Ocean, the statement confirmed his worst fears about Sea Shepherd.

Watson was disgusted and energized. He lived for this. It was 1010 on Christmas eve, the *Farley* was being battered by a building antarctic storm, and it was time to fire off some squibs and see who he could piss off more. He lumbered back to his radio room, hardly noticing the hard rolls; sat down at the keyboard and composed a rebuttal to Campbell, saying that the Australian navy is a dog being kept on a chain while the government helps the burglars. "The [Australian] government says that the Japanese do not recognize the Australian claim to the Antarctic Treaty," he wrote. "In 1942 they did not recognize Australia's claim to Australia."

In 1988, Watson flew back to Iceland to face charges for the sinking of the two whaling ships. He egged the prosecutors on, then practically ordered them to charge him. The poor, dour, polite Icelanders—they had no idea how to deal with this cagey bear. The lead prosecutor, growing ever more suspicious of being led by the nose into some trap, finally flashed, "You can't make us prosecute you! This is our case! You said you weren't actually at the sinkings? Well, then! You are not guilty." Watson was released the next day. According to Watson, the Icelandic minister of justice said, "Who does he think he is, coming to Iceland and demanding to be arrested? Get him out of here." The Icelanders breathed a sigh of relief when they got Paul Watson out of the country.

"Here's the other thing," he said, looking up from his keyboard, "This is a Canadian ship. The Australians have already said they're not going to enforce international laws in their territory. So this is between the Canadian authorities and Sea Shepherd."

The last section of the article Watson had just read from said that the Japanese crew member with appendicitis had been winched aboard a rescue helicopter from the spotter vessel *Kaiko Maru* about fifty nautical miles south of Tasman Island. Plans for the ship to refuel in Hobart had been scrapped in the face of demands that the ship be restrained to keep it from returning to the whaling grounds. After delivering the sick man, the ship had sailed south again.

Watson's media campaign was having some effect. He and Greenpeace had focused so much negative attention on the Japanese fleet that they could no longer safely enter an Australian port.

At 1145 Watson told Mia to change the course to 280 degrees. This would set us directly downwind, running with the seas, and thereby eliminate much of the hard roll. Now the *Farley* rolled only when she surfed down the front of one of the giant waves and yawed. The new course would also send the *Farely* safely south of the last known Japanese position. Watson and his officers knew that the whaling fleet would have to head south again at some point, and there would be less risk of missing it again if we were ahead of it and waiting.

At 1200 we all went to lunch. Running with the seas was theoretically the least jarring ride. Try telling that to your pea soup. Every fourth mouthful, the ship would surf down the front of one of the building waves and yaw off course, and cant over, and eighty-eight hands from one end of the mess to the other, in the engine room, and no doubt, in the cabins and bunks, would find themselves in a sudden dilemma: grasp for the tabletop, bulkhead, or bunk edge and save themselves, or steady and cover the soup, the spanner wrench, the pen and journal. The results of many of these choices were revealed in the next instant when a coffee mug, a plate of pasta, and George the videographer went flying across the mess.

The barometer on my altimeter watch was dropping like a stone. Generally, the lower the pressure, the stronger the storm. There was a little graph, courtesy of Casio, that showed the barometric pressure in three-hour increments; it looked like a staircase to hell. It now read 29.10. How could it be dropping this steeply now, when we had already been in a building gale for two days? The wind in the last twelve hours had also swung around more to the east from the southeast. When it had howled from a more southerly direction, it had essentially poured off the continent of Antarctica. There were mountains, jutting peninsulas, and lots of big ice—all sorts of obstacles to obstruct and break up the wind. But look due east of the *Farley* and all you saw was ocean. Unobstructed Southern Ocean girdling the naked globe. The two variables that govern the wind's effect on water are wind speed and reach, reach being the distance the wind travels unobstructed. Here, around the disk of Antarctica, for something like 10,000 miles, there was nothing but reach.

The *Farley* pitched forward, yawed to port, and gouged a white tear out of the trough. Down in the bottom, we were surrounded by walls of water. The backside of the wave ahead was monstrous and easily blocked our view of the horizon, putting it at about twenty-five feet. I went to the hatch to the port bridge wing, looked aft through the thick window, and saw that the next wave barreling in behind us rose black and ominous over the level of the high heli deck.

The whole bridge shuddered in the heavier gusts. Watson said over the booming, "Ever see the bridge set of the *Enterprise*? On *Star Trek*? It's on springs—they shake it. And steam. Are they running on *steam*? Every time a pipe breaks, steam comes out. And flames come out of the control panel. You'd think by then they'd have better technology."

Many of the rest of the crew were lying low, trying to catch up on sleep, disheartened by the near misses. The frenetic preparations had ceased. The pressure drop of the storm mirrored the mood, the depressing suspicion that we might be on a wild goose chase. The ship was quiet, if a ship in a gale can be said to be quiet: the hull shuddered, sloshed, and boomed; the engine throbbed and growled; provisions and anchor chain clattered and thumped in the holds; the steel bones of the *Farley* groaned. Anything slung on a hook knocked against the bulkheads. And wind. It drove the ocean westward like a herd of terrified cattle in which the *Farley* floundered like a blinded horse. The ship was concentrating all of her energy outward, trying to keep her balance and her speed, and inside a stillness took hold. Nobody ran the grinder over newly fashioned weapons or bragged about prop foulers. The companionways were empty.

Alex should have been asleep, but he came onto the bridge with Aultman. Hammarstedt was already there, as it was his watch. It seemed that in a storm this big, none of the officers wanted to be too far from the bridge. Watson said to his first mate, "I did get a message from M——. They're zigzagging all over the place, tending south." He nodded to Aultman. "I think once we get down there where it's nice and calm, we'll be sending the helicopter up to find them along the coast. I think they're just trying to shake everybody off. Maybe they're just waiting for the other boat to rejoin them. They've got to start whaling sometime. The other boat didn't refuel in Hobart. They stayed outside the port. So they might be waiting for them to refuel at sea."

Watson turned to Hammarstedt. "What's the longitude of Tasmania? What's their position relative to Tasmania?" Hammarstedt held up a hand for Tasmania, and pointed with a finger for the fleet. "OK, so they're pretty much right below Tasmania now; maybe they're just waiting for them."

"Just over 1,000 miles from Tasmania," Hammarstedt said.

"About three and a half days, I guess," Watson said. "I have a hunch they'll be heading to Porpoise Bay next."

At 1430 we were at 63°05'S, 142°18'E, heading west with the wind and the seas. The captain had finally retired for a while. We fell into a trough, and the world was suddenly circumscribed by two walls of dark water—one forward, one aft. They must have been thirty feet. The top of the one ahead of us blew off white like a spuming snow cornice. I looked off to starboard and a wandering albatross, nearly as white as the froth, wings motionless, slipped down the face astern and glided up the face ahead, just as we pitched upward and began our own floundering rise. The albatross hit the top and canted her soft belly to the storm, and made a screaming banked peel-out downwind and over the other side. I don't know if anyone else on the ship saw her. To me, she was a visitation. Not harbinger or annunciation, but a simple reminder of a world that worked, that was at home with itself and friends with storm.

Hammarstedt said that late last night, on his previous watch, eight humpbacks swam by and surrounded the ship.

The barometer was still in a nosedive. At 1600, when Alex came back up for his watch, it read 29.05 and dropping. He immediately told Gunter to swing the ship around to 234 degrees, and he swapped out the two wooden number blocks on the console. The three yellow rubber ducks were up there, faring resolutely forward, and he took them, too, jamming them next to the Hayagriva to keep him company. Alex was all business. He showed me the chart.

"We can either mess around here or head here—" sixty miles north of the Dibble Ice Tongue, on the coast—"which is what the captain decided to do, and I'm glad. We'll arrive there in twenty-two hours. Sit next to the ice and wait for them. If they go east we can move east; if they go to Porpoise Bay we can trap them in the bay. Going up here, it's just getting rougher and rougher. Can't launch the helicopter, can't do nothing." That would put us there at 1400 on Christmas day.

The *Farley* was moving forward, but the waves were moving faster, so that it felt when they passed as if the ship was skidding down the back sides of the swells as they barreled through. The *Farley* pitched down a bigger set wave as Gunter swung her to port, and she yawed and canted over and when the next wave hit her on the stern quarter, she rolled thirty degrees. It steadied to the new course, rolling heavily and hard, and five minutes later another set wave came through and the *Farley* slammed over and I heard more miscellany hitting the bulkheads and saw the wave break over the bow.

Aultman grabbed at the bolted-down stool on the port side of the bridge, and Ron grabbed Aultman and his video camera at the same time. "Now *that* was forty! At least," shouted the pilot.

Dinner was more like an athletic exercise than a meal. When Wat-

son went up for his watch afterward, he tried to turn the ship back
into the seas to take out the roll. A weird tearing fog had set in, and the
ocean was marbled white. The whitecaps exploded and streamed the
dark sea with spittled strings of foam. The *Farley* was slamming head-
on into the waves that were cresting at thirty feet. They broke over the
bow and washed the main deck, and there was no horizon; there was
only the deep trough, and black and white water no longer tossing but
heaving with a monolithic power, a three-story wall held up against
the sky by an accumulation of violent forces—days and days of them,
thousands of miles. That's what met the *Farley*'s bludgeon bow, and
washed over it with a flood of green water, and then the ship reared
like a terrified horse. The wild launching of the bow into a fog-
spumed featureless sky—that was the place where time stopped.
Where we held weightless, not of water or earth or air, and the longer
the moment, the more likely the sawing descent would be a jarring
slam, an explosion of white out of the next trough. Any big chunk of
ice could have torn her open. I knew, watching the ship taking wave
after wave, that in this water—Trevor had told me this morning that
the intake pipes were reading thirty degrees—a man overboard, or
ship damage necessitating the launching of lifeboats: well. The
lifeboat would never get launched; the man would never be recov-
ered. If something bad happened out here there would be no rescue.
The safety gear we had on board wouldn't do a bit of good.

At 1130 Aultman staggered up to the bridge in his Mustang suit
and went out through the aft hatch. He took Casson as a spotter, and
they were harnessed and tethered. When he came back ten minutes
later he told the captain to turn the ship around. Running into the
wind was shaking the rotors of the helicopter apart, bending them
nearly to the deck as they pulled against the tethers at their tips. Along
that axis, the composite blades had very little strength and could snap.
The wind might have been sixty knots. Turn it around, he begged, or
we'll lose the bird. Watson said OK. He ran west with the storm. It
was a significant decision.

15

Force 8

It was three o'clock when I awoke on Christmas morning. What woke me was the sudden drop of the bow as it gashed into the trough, and the impact of my right shoulder hitting the locker at the head of my bunk. The *Farley* shivered. Then the wave pitched the stern out of water and the prop howled, beating air. Water gurgled along the hull and sloshed at the bolted porthole. I lay for a moment and breathed, and listened. Something was different. Something in the pulse of the ship, a faster beat of the engines, a quickening. I swung out of the bunk, grabbed my dry suit and the life jacket I had bought in Hobart, and ran up the three lurching flights of stairs to the bridge.

The sea was a frenzy. No night's respite, no night at all, just the unabated gloom of a perpetual dawn. The waves were now over thirty feet, and the wind tore off their tops and streamed their backs with ropy lines of foam. Snow blew by in the tortured fog and mixed with the plumes of exploding spray. The *Farley* took the monsters under her stern, and the exposed prop shuddered for a moment like a thing in pain before the wave threw the bow wildly to the sky. It was now a full force 8 gale. The timbers of the bridge creaked and groaned and the wind battered against the superstructure and against the half-inch Plexiglas windows with a pitched moan like an animal.

Watson sat up in the high captain's chair in his Mustang suit and Sorrel boots, looking alternately at his radar screen and at the sea. He was focused and calm. Alex was in the center of the bridge at the helm, trying to keep the *Farley* running straight on the waves. Hammarstedt was at the main radar to his left.

Alex said, "Good timing." His eyes were red-rimmed. "Two ships on the radar. The closest is under the two-mile range. If they're icebergs they're doing six knots."

"Probably the *Nisshin Maru* and the *Esperanza*," Watson said. "They're just riding out the storm." Where the *Arctic Sunrise* and the five other boats of the whaling fleet had scattered in the gale no one could say.

I stared at the throbbing green blips on the main radar screen. Was it possible? Had Watson found, in hundreds of thousands of square miles of Southern Ocean, his prey? It was against all odds. Even with the informer on board the *Esperanza*. Even with the storm that could now be veiling his approach from the unwary Japanese. I looked at Watson in his exposure suit and began to pull on my own dry suit. Watson turned to Cornelissen. "Wake all hands," he said.

We were 220 miles north-northeast of Antarctica's Commonwealth Bay, about twenty miles west of where we'd been when I went to sleep a few hours before. The *Farley* labored up the back of a thirty-five-foot wave and plunged down the other side. Green water poured over the bow and flew up in a white explosion that battered the windows. We were running with the gale. It howled out of the east-southeast. We had not gotten an update from the mole on the *Esperanza* since yesterday afternoon, so if this really was the *Nisshin*, we had been brought directly to it by the storm.

An intense quietness had come over the bridge. After many months of preparation and planning, Watson was sneaking up on two vessels in a vast empty sea in a near hurricane and the radio was silent.

No one spoke. Allison came onto the bridge, as did Aultman and Ron with his camera.

At 0350, through the fog and spray, maybe three miles ahead and a little off to port, I saw a shadow that was not mist or wave. It was a ship.

Alex said, looking over Hammarstedt's shoulder, and steering with the manual control, "We got three targets. The second ship is two miles ahead of this one."

Watson said, "The only three boats would be the killing boats. Christmas day would be a good day to grab them. Don't know what we can do out here, but we can do something. Get Price up in the ultralight." He was joking.

"Looks like the *Esperanza*," I said.

"Yeah, it does, sort of," said Hammarstedt.

"The bigger one beyond must be the *Nisshin Maru*."

Watson said, "If that's the *Esperanza*, that's the whole damn whaling fleet. Maybe the whalers think we're the *Arctic Sunrise*."

"You wanna sneak behind the whaling ship or behind the *Esperanza*?" Alex asked.

"Go right by Greenpeace."

Silence. Everybody was straining to see through the snow and fog and blowing water. Watson said, "We've got the advantage in this kind of storm because the ship was built for this weather. How far to the nearest Japanese boat?"

"Two point seventy-five miles," Alex said. "The *Esperanza* is half a mile. The big ship is moving about seven knots."

I looked up at the GPS, and we were running ten knots, going with the storm. We were close.

The *Esperanza* took shape out of the mist. It was much bigger than the *Farley*, gaily painted blue-and-white with a big rainbow arcing across its side. It had white satellite and radar spheres stuck to the top of the superstructure, and a fancy enclosed crow's nest with canted windows, and an open heli deck aft with an enclosed hangar that must have made Aultman green. The *Farley* came up on its stern, a black apparition, and passed it.

"That's a $30 million boat," Watson said. "Just the retrofit."

"I've always wanted to overtake a Greenpeace vessel," Alex said. "Nice heli deck."

"There's so much we could do with a ship like that," Hammarstedt murmured.

And then we noticed something astounding. Its starboard bridge wing was crowded with people, and they were waving wildly and pumping their fists, and one was swinging a big Canadian flag back and forth. I blinked. They were saying, Thank God you're here—stop the bastards.

We *were* all in this together. They had been harassing the whaling fleet for days now, watching whales harpooned and exploded, electrocuted and drowned close at hand. They had been doing what they were mandated to do—letting the world know it was going on. They waved us on. A minute more and the Greenpeacers on the deck were swallowed in mist and spray and then the ship itself was again a pale shadow. And then we were thrown over to port and two bodies went flying across the bridge, and a collective gasp went up as the rest scrabbled to hold on to something.

"That's a forty-degree roll," Alex said.

"That was at least a forty," Aultman said as he straightened.

The *Farley* seemed to shake herself off, and resumed her patient climb and plunge.

"Valley Forge was on Christmas day," Watson said.

"The *Nisshin Maru* is increasing speed—*Nisshin Maru* 6.8 knots, we're doing 9.4."

"I don't think they can go any faster. They're riding it out."

All eyes ahead. We plunged off another wave, yawed. From the top of the next, I made out a thickening of mist, specter-like and monstrous in size.

"There it is!" Hammarstedt cried. "That's definitely it!"

"Oh, yeah, that's it." Alex. He was at the manual controls, flicking his eyes up to the GPS for speed and bearing.

A terrible suggestion of a ship. Straining, as if the eyes themselves could concentrate and clarify the image. Then the fog did it for us, rending like ripped gauze, and there for a second was the giant stern, the slipway, the white block letters that read *NISSHIN MARU*, TOKYO.

Trevor, with his engineer's telepathy, down in the engine room, must have known. The *Farley* was taking the waves at what for her was a dead run.

Alex said, "We're doing eleven knots. They're doing six."

"I have tweaked the engine." Trevor was in the doorway, ear protectors propped on his head, an elusive smile just slipping away.

Allison said, "The captain says we're severely limited with what we can do in this weather. Don't want anybody out there—hell!" She pointed out the forward windows. "Look at Gedden out there!"

Gedden, the tree climber, was crouching, moving forward on the forecastle like a man in battle under fire. He had something black clutched against his right side, and when the bow plunged he grabbed the water cannon, or the anchor chain. He scrambled forward now on hands and knees. An explosion of spray covered him and he moved forward again. The next one knocked him over. When he got to the bow rail he whipped a carabiner from his harness and clipped it to the rail. Tethered in, holding the thing against him with his elbow, he went for the flag mast at the bow. He was going to hoist the Jolly Roger. In sixty-knot gusts. He did. He got the flag clipped on somehow and hoisted it and cleated it off. Then he waited for the first climb out of a plunge and unclipped himself and scrambled aft.

This was a psychological war as much as anything. The Japanese had said in their press when they left on November 7 from Shinoseki harbor that they were afraid of an attack by Sea Shepherd. Fear would make them run. When they ran, they did not kill whales.

I didn't think fog happened in near-hurricane winds, but there it was. Shrouding the *Nisshin Maru* after the first glimpse. The ship might have seen us, but probably it hadn't. It was maintaining its speed, just under seven knots. It was a sitting duck. It thought we were the *Arctic Sunrise*. Nobody bothered to step out on its bridge wing to look back and check. I could only imagine what we would look like ourselves, appearing out of the ragged fog, black and battered, the gale-stiffened Jolly Roger flying, with an avowed mission to cripple or destroy.

Alex kept one eye on the radar now and one eye on the sea ahead. He had the ship targeted on the screen, in a small white box, which gave him a continual readout of its speed, direction, range, and time to contact.

"Seventeen minutes," he said. "Twelve knots."

Trevor murmured, "Nice."

Gedden stepped dripping onto the bridge. "Kind of limited with the Zodiacs," he said, catching his breath.

"No, we're pretty much the can opener right now," Trevor said.

We took another wave over the bow, green water, and I thought that if Gedden had been out there then it would have probably knocked him over the rail and tested the strength of his tether. Another wave flew up hard against the windows, and then it was as if we had come through a curtain. The fog ripped away and just ahead was the slipway ramp cut into the stern where they winched up the dead whales, and the tall white superstructure of the cranes. A banner over the slipway read, "Greenpeace Misleads You." Fighting fire with fire. Running down the length of the hull, visible when it corkscrewed on a swell, was the large block-lettered word RESEARCH. As the ship rode over the bigger waves, its prop came out of the water. We were 1.2 miles away.

Watson said, "I think the best tactic here, Alex, is the prop foulers. Bring it as close to the bow as possible. Low profile as possible. We don't want them to see what we're doing."

Alex was leaning forward into the window in front of him like a cat watching a chipmunk, except that the chipmunk was the size of a city block. His eyes didn't move off the stern. "Do we want to ram them? Punch a few holes in their ship?"

I thought it was a rhetorical question.

"No, we'd sustain a lot of damage. Prop fouler's the best thing right now." He seemed to be protecting his crew. No sane person wanted a collision in these seas. Watson turned to Trevor. "Tell them to get the prop foulers ready on the stern. Tell them to stay down, stay hidden. Don't deploy them until I blow the horn."

Trevor nodded, exited. Allison tugged at Watson's elbow. "Do you want to get the helmets and vests? This is where they're going to shoot."

Helmets and body armor. Must be some of the items behind the locked door labeled "Powder Room."

"Sink right to the bottom with that stuff," Alex said. "I'll take a few bullets."

"Yes, but the bridge is where they're going to shoot."

"They're not going to shoot us."

"I can't believe they're going so slow," Watson said. We were coming right up on their stern, half a mile now and closing. "How far is the *Esperanza* behind us?" The proximity of a potential rescue ship in these seas might determine Watson's level of aggression. Though I didn't think a rescue boat could even spot swimmers in the violent waves.

"One point two miles," Alex said. "We've got a third ship straight to port six miles. They're doing only five knots."

The *Nisshin* was very close, but not close enough. It was like sneaking up on a browsing deer, holding your breath, praying a twig wouldn't snap. Maybe somebody on the *Nisshin* finally saw us. The pace with which we closed the gap slowed a little.

"The factory ship is increasing speed to eight knots now!" Alex said.

It was 0448. Alex began ticking off the *Nisshin*'s speed. "Increasing speed again, nine knots now. 9.4. One knot more."

Watson watched his prey. He sat up in the captain's chair, one hand, out of habit, on the knob of the lever that controlled our speed. He worked his jaw to the side. Not excited, not angry, just focused. He did look like a polar bear, with the same pitiless detachment, the sense one got of icy calculation, weighing distances, speed, odds. It was easy to see that this was not his first rodeo.

He said, "Tell Trevor to deploy at the last moment. We don't want them to see us put it out."

Alex said, "Nine point five."

Watson didn't take his eyes off the ship, "I don't think they can go much faster in this weather. That's where we have the advantage. Some horses are good in mud and some aren't. This is a North Atlantic trawler—that's where our advantage is."

"Our speed is eleven," Alex said. "Their speed is nine and a half."

"Where's the harpoon boat?"

"Falling back six miles."

We were off to its starboard, coming up on its stern. It was monstrous. Even so, the bigger waves were throwing its prop out of water. Some of them were over forty feet. Trevor shoved open the bridge wing door and entered. He had on his Mustang suit now, and he was wet.

"The horn is on. They have the trail line ready." The trail line was three-quarter-inch thick longline on a spool on the stern. There was half a mile of the stuff. Watson would try to push across the *Nisshin's* bow while Trevor and his team unleashed the floating line. The bigger ship would have no choice but to plow over it. The line would work its way down the hull and, Watson hoped, get sucked up in the *Nisshin's* prop. But I didn't see how that would not be dangerous to the Japanese crew in this kind of storm. If the prop did jam, the *Nisshin* would have no headway and would broach sideways to the seas and wallow. They'd have to launch life boats in seas like this, in water that was at or below freezing.

Watson said again, "As soon as I hit the horn, then deploy it. We have to get far enough ahead so we don't hit them, but close enough so we're effective."

"Nine point six."

"Keep pulling alongside."

"I'll let you be the judge how close to get."

"We'll lose some speed as we cross."

Watson said, "I hope we can disable those bastards. They're not going to let us have a second chance. Where are the passports?"

Kalifi said, "In the safe."

"The bag with the passports is not in there. Get someone to find them."

Kalifi went out. Watson was making preparations to abandon ship if necessary.

Alex, his voice rising, called out, "We're getting pretty close here. Point four miles."

"I'll hit the horn when I want Trevor to deploy."

"I found the passports."

"Just put it by the safe so it's ready to go."

"Greenpeace is speeding up too," Alex said. They had to be glued to their radar, watching the signal blips of the two ships starting to overlap.

"We could ram her up the slipstream if you want," Alex said. "What do you say, Paul?"

"Yes, the swells are good," piped up Hammarstedt, who was again at the main radar; it was the first thing he'd said in a while. He meant that we had following waves of great size, so it would be easy to come down off the top of one and crack down with force into the opening of their slipway.

"No, we're going to do this," Watson said.

Watson had the benefit of three decades of actions similar to this. He was evidently making his bearlike calculations.

"Is he picking up speed at all?" he asked.

"Nine point nine knots."

"He can't go any faster. He's going to cut the swell as soon as we go by him, too." Meaning Watson thought he would turn.

"Toward us?" Alex asked.

"Yes. Don't underestimate the guy gunning it as we come across." Watson was thinking they might try to ram us. "I think just past the 'Research' thing is the best time. How far in front of the Research sign do you want to start cutting in? You feel safe enough?"

"We've got a knot," Alex said.

The *Farley*, to everyone's astonishment, overtook the *Nisshin*'s stern and began to move up alongside. It was about 300 feet to port. Black hull, white superstructure, freshly painted, four stories high, with three massive crane gantries whose tops must have been seventy-five feet off the water. RESEARCH. Clean, innocent block letters. We were edging up along the word. When the bridge reached the H, Alex would swing in toward the ship. He would count on our extra speed to angle us in front of the bow and get us clear across it before a collision.

"There's nobody there, nobody even looking at us," Watson said. "I think we caught them unawares! There's nobody looking at us."

Just then it was as if the Japanese woke up. It was as if the *Nisshin Maru* jumped in surprise. Someone put the hammer down and it began to pull away.

"He's turning away," Alex said.

Watson, curt: "Turn with him—is he speeding up?"

"Turning away—not very smart, they're going to take the waves on the beam—Yes, they're getting away! Ten point seven—matching speed—eleven point three—they're faster, eleven point five, eleven point seven.

"Go right on their ass, then—"

As Watson ordered it, as Alex began the turn to follow, a hard wave hit the *Farley* on her port quarter and she slammed over. The waves weren't getting any smaller. Some of them had to be forty-five feet. We were falling back along *Nisshin*'s endless aft deck. And we were now taking the seas, like it, on our port stern. It was not a good angle to the storm.

"Go for it if you can. Straight into the slipway."

"Oh, fuck, yeah," said Alex. "I hope they stop, got a surprise for them. They are a little faster, but not much: eleven point five."

It was too late. Trevor had tweaked the engines, and the *Farley* was straining with all she had, eleven, eleven point six, twelve knots. But the *Nisshin* was too powerful. She came up to speed and fled at sixteen knots.

And then it was as if the *Nisshin*'s skipper snapped. Captain D. Toyama had been whaling down here for decades. He had been harassed for days by Greenpeace; their Zodiacs swarmed his killer boats, his harpooneers had shot whales right over their heads. And here, out of the fog, was a ship willing to disable his own. He'd had enough. The *Nisshin* was a quarter-mile away when it turned to starboard, angling across our bow and slowing down. Toyama seemed to be saying, "OK, you want to mess with me? Come ahead."

Alex matched the turn, all but thirty degrees of it, so as not to fall behind the *Nisshin*'s stern, and set a collision course. He too was completely calm. Watson, out of his chair now, stood with a hand on the lever that controlled our speed. Now we caught the crossing seas on our starboard and the *Farley* slammed over to port in a forty-degree roll that sent Kristian crashing across the bridge. The *Farley* righted and slammed to the other side. Alex looked at the radar. He turned to Kalifi and said, "Tell the crew collision in two minutes."

Most of the crew were gathered in the mess in their exposure suits, aft of amidships, below the deck, and a long companionway away from the main hatch exit. One of the officers—it wasn't clear who—had ordered them there. Not a good place to be. If the *Farley* broke apart they wouldn't have a chance of getting out.

The *Nisshin Maru* was on our port side, and the two ships approached each other at an acute angle. By the law of the sea, in a collision situation, we had the right of way, as we were on his starboard. The *Nisshin*'s bow, as tall as a three-story building, lunged off a thirty-five-foot wave, airborne, and crashed down like a giant ax. The hole it tore out of the sea vaporized and was driven downwind. The gap between us closed: 300 yards, 200. Now we could hear the blare of their horn through the tearing wind. Repeated blasts, short and long, enraged.

"Collision one minute."

I tugged on the waterproof zipper of my dry suit and had one thought: you're going to be wet and cold in about twenty seconds. The hammering bow loomed, 200 feet away, aimed at our belly, amidships.

Alex glanced at the radar, at the juggernaut, held his course. He was focused, intent. A deadly game of antarctic chicken: 150 feet away. Alex blew the horn, which was the order to unleash the prop fouler. A squad on the stern stood, braced themselves, and whipped several hundred feet of the mooring line off a big spool, enough to tangle any propeller.

And then the *Nisshin* blinked. Whoever was at the helm threw it over to port. For an agonizing second the two ships ran parallel, and then they were pulling away. They quickly resumed full speed and fled back into the fog. As they ran, Watson pulled down the mike on maritime channel 16, and barked, "*Nisshin Maru, Nisshin Maru,* this is the *Farley Mowat.* You are in violation of an international whale sanctuary. We advise you to get out. . . . Time to go now, you murdering scumbags. Now move it! And run like the cowards you are."

I looked at my watch: five-forty-two a.m.

Watson handed the mic to Casson, who spoke rudimentary Japanese. "*Nisshin Maru,*" he said. "*Nisshin Maru,* you are murderers. You are dishonorable."

Alex lifted his watery blue eyes from the radar and smiled. The first one I'd seen in days. "They actually increased speed when you said that."

"Go back to Tokyo," Casson said. "Good luck."

Everybody breathed.

I asked Alex if we would have sustained damage at those speeds. "They would have sunk us," he said. He handed over the helm to Gedden and moved aft to the chart room with Watson.

I asked him again, "You're sure?"

Alex is usually in motion, but one of the things I liked about him was that when he stopped, he gave you all his attention. "A ship that's ten times as heavy as your own ship—" he said. "That hits you amidships with its bow, its gonna basically slice your ship in half. That ship will completely destroy your ship in a matter of seconds. Everybody who would've been inside would have had a very hard time getting out."

I nodded. "There was a point there where it was up to him to whether we were T-boned or not."

"Yes, he definitely had that choice, and he didn't take it. If he would've ended it there, that would've probably ended commercial whaling. But I still believe that not sacrificing people for that, in that way, is probably a better choice."

"But personally, you're willing to make that trade-off—trade your own life for stopping whaling?"

"Absolutely. But I'm not going to engage in some suicide mission. It's gotta be a calculated risk."

The captain said, "If they had sunk us, there'd be such bad PR for them. The Australian navy would be down here in no time. They'd be hauled in for investigations. Australia would have to intervene at that point. We have Australian citizens on board."

"But no one would have seen," I said.

"The *Esperanza* was close behind. Close enough to see. Legally, the *Nisshin Maru* had to give way. I think he might have hit us, but he could have shut himself down. If he hit us in the bow he wouldn't have sunk us; he'd have to hit us amidships to sink us. But he would sustain so much damage he'd have to go home. And if he ran right over us his propeller would be mangled."

Watson looked doggedly out from under his shag. "We've always won every game of chicken we've played," he said. "We did it with the Spanish navy, bow to bow. I think they were full speed. Their ship was 100 times what ours was worth. We just kept going. We knew they'd turn off and they did. Did the same with the Russians. I told them, 'Your ship is worth a hell of a lot more than ours is, so you'd better get out of the way—'and they did. Well, we're chasing after them. I wonder where the harpoon vessels are?"

Watson ducked into the radio room. By 0605 he already had his first press release posted. It began: "No whale will be killed this Christmas day . . ."

I peeled off the dry suit and let the adrenaline wash through me. I thought, No doubt now—Watson is surely an anti-Ahab. More bearish, more charming, but just as terrifying in his fearlessness, and in his willingness to sacrifice everything, including our lives—to save the whale.

A certain somberness took over the ship. The storm raged. Watson showed me the weather fax. "Looks like a freight train," he chuckled. "Never really seen one like it." A line of five L's for "Low"—tightly packed storm systems—marched one after another east to west across the sixty-second parallel. We were at south 62.59 degrees, in the middle of the track. We couldn't turn south, because we'd take the waves on the beam or the quarter and the old *Farley* would be pummeled. Two forty-degree rolls had been violent enough. Minutes or hours of them would shake the ship apart. We had to run with the storm, and that is what Alex did. No one wanted to be in the middle of it, but there was no choice.

Alex set the *Farley* down the waves with a heading of 290 degrees WNW, the exact bearing of the *Nisshin*, which, even with its size, evidently didn't wish to challenge the gale either. After getting some distance between itself and the *Farley*, it had settled back to a more reasonable fourteen knots and was patiently opening the gap. The *Esperanza* had set the same course and was moving up on us and would soon pass. Alex tracked them both on the radar. Greenpeace was fast on the media blitz, too. They posted a blog almost immediately that said they were in a force 10 or 11 gale and that the Japanese weren't hunting on Christmas because of the severe weather, and included a one-line mention that Sea Shepherd had arrived in the area.

I went down to the mess. It was just after 0700, an hour and a half after the near collision. The halls were empty and almost no one was in the mess. Wessel and Geert were there in a booth. Wessel had been out on the stern, exposed in the storm, one of the handful of crew who had run out the fouling line. He was out of his Mustang suit now. He looked a bit dazed. I slid into the bench seat opposite.

"I thought this is it," Wessel said. "I was ready to die there. But I had a good feeling in my heart. I was watching some of the guys on their ship. I was holding on for dear life. When they turned toward us, and we turned just a little off the waves, the wind tore into the aft deck and me and Luke went flying. I grabbed hold of a pole. Wind and rain so slippery—my grip was loosening—a forty-degree roll—I thought, Jeez, I'm going overboard. Just then Luke grabbed my headgear and held me, and I stopped and braced myself on the deck. And I looked up and saw the *Nisshin Maru* and saw four or five Japanese people looking down at us with arms crossed—they looked quite detached. Just then the *Farley Mowat*'s horn blew. And that was the signal for us to let out the thick blue line rolled on the roller. We let the whole thing run out and then we cut the line. But I think they saw us do this. Right after we did that, we got ready to launch a steel cable prop fouler—as soon as we were ready to drop it in the water their ship turned. And thank God, because thirty seconds later we would have collided. One hundred yards away, they were taking pictures of us. I think they saw the blue line and that's what stopped them. I think they were bluffing, but they gave us good warning, they were blowing the horn for quite a long time. Exactly what we wanted—now they're running with their tails between their legs."

Within the last couple of hours he had nearly died. "Only one drop of pee," he added and grinned.

Geert said, "From the back you can see the rudder coming out all the way—could see the whole bloody thing."

"It was fantastic." Wessel took a deep breath and put his back against the bulkhead. "That was a defining moment in my life, I reckon. From now on I'm devoting my life to saving the whales, direct action. I think a lot of people on the ship feel the same way—we were dedicated, but that sort of settled it for me." He shook his head. The images of his near demise kept intruding. "How one of us didn't fall in. Just amazing. It's not that we were unprepared, but things happen so fast—rain and snow so slippery—one little shift of the boat and you're on your ass sliding." He shivered.

I climbed back up to the bridge. On the way up the narrow main companionway I ran into young Willie Tatters coming out of the hatch to the engine room. His dreads were stiffened with engine oil, his over-the-knee shorts were torn and he looked generally smoked.

I said, "Good job, Willie."

He paused, looked over his shoulder, mystified.

"What do you mean?"

"Were you down with the engine for the ramming?"

He nodded, unsure where this was going.

"Good job, we were going fast."

It was like pulling a fish taut on the line. He swiveled. "What was the speed up to?"

"Eleven point five knots. I think twelve at one point."

His whole face brightened through the grime. He smiled to himself, nodding. "Whoo-hoo," he murmured and continued on.

I got to the bridge in time to see the *Esperanza* passing us with its festive rainbow paint job, and to hear channel 16 cackle with the voice of its captain.

"Captain of *Esperanza* to captain of *Farley Mowat*: the captain of the *Nisshin Maru* will ram you if he gets the chance. Good luck."

Gee, thanks, I thought. The warning's a bit late. Toyama had the chance and demurred at the last second. Or maybe one of his officers grabbed the helm. Though at the speed the *Nisshin* was traveling, as close as it came in those seas, one small miscalculation would have sent its bow ripping into the *Farley*. Alex had repeated with a kind of awe how hard it was to steer the ship in the quartering waves. With the *Esperanza* tucked safely back two and a half miles, any swimmers would have surely been lost.

I was amused to see camera flashes popping from the *Esperanza*'s bridge wing. Its crew were taking pictures of us. The Jolly Roger was flying, stiff as a board, and the edge of the flag frayed in the punishing wind. The *Farley* tore white out of the ocean with every plunge of her bow. She must have looked mean.

It was 0748. Twelve more minutes of Alex's eventful watch. He had peeled off his exposure suit and was still at the helm, guiding us down the waves.

He was still mulling how close he had come to cracking open the *Nisshin* like a fat nut. "We're pretty much the same size as the chasers. When the ship went down and we go up on a wave, we can ram right up in the spillway. Probably get stuck in there, which is perfect. We can cause some damage, send them home. If we could have gotten closer I would've rammed up their spillway, absolutely." He smiled his short, tight, efficient smile. "Timing's gotta be perfect. We can get the bow in there. It's a matter of aiming. It's a bit rough in this sea."

"Do you think the Japanese are raising alarms now with the Australians, with their own government?"

"We didn't do anything. We were just towing a line behind us with laundry on it." He flicked his eyes from the windows, where thick snow blew by, to the radar, and pointed at the blip that was the *Nisshin*, ten miles away now. "I guess you're right. I wouldn't be surprised if they already called the environment minister, Ian Campbell."

Then, very matter-of-fact: "I was surprised they didn't ram us."

It was like saying he had actually expected most of us to die.

"But if they had, you said we would have sunk."

He smiled. "Oh, definitely. Did you see the size of that thing? Would've cut us in half. In sea like this I'd prefer to keep her afloat—"

No kidding.

"—If you can't get into the life raft you're fucked. I just wanted to get in front and past. When he came toward us I moved away slightly. I was going to cut in at that point and ram, but he already turned away from us. We needed a little more speed. Two more knots would've done it. I only changed like ten degrees; it was no big deal. Fucking hard to turn in these seas, man. Fucking hard."

Alex had been doing everything he could to cut in front of the monster and deploy the prop fouling line where it had the best chance of doing the damage.

"If he would have made the slightest error, we would have been in front of him and he would have hit. That's why I told everybody— wear your Mustang suits. In this water. Greenpeace would have loved to pick us up. Good for us, too."

Except that most of us would have been lost. We both looked out the shuddering window at the horizontal snow, and at the spuming waves, which even now were bigger, more violent than they had been an hour ago.

Alex glanced at the clock on the GPS screen. "Hope that storm dies out in the next twelve hours so we can send out the helicopter, keep track of them. Greenpeace is keeping up with them. Seems to me we are tossing the ball here. First Greenpeace chases them away. Then we chase them away."

The watches changed. Mia and Lincoln took over while Watson hammered out more PR in the radio room. The sultry French girl looked dreamy. She sat up in the captain's chair cupping a lidded mug of coffee while Lincoln steered, her dark hair spilling around her face onto her delicate shoulders. "I want to die with a harpoon in my chest at eighty-six," she said as spray hit the windows. "That would be a beautiful way to die. Of course I hope there is no whaling by then."

Watson entered the bridge, reached up, and plucked the mic out of its cradle. Mia slipped out of the captain's chair without being asked. Watson hailed the captain of the *Esperanza*. He thanked him for the call earlier. "Captain Boyer, you're doing a good job. I don't think they'll be killing whales today," he said. "They're pretty skilled. Yesterday was rough as well; they still happened to kill whales." He released his thumb from the transmit button.

"The captain of the *Nisshin Maru* is a mad dog and won't hesitate to ram you."

Watson grinned.

"The killer boat captains are more polite," Captain Boyer continued. "Yes, they killed five whales yesterday, even in this rough weather."

"Have a good rest of the day. I'm sure we'll see you."

All very cordial. Why shouldn't they be? Watson's beef was not with the crew, but with the brass. Then more crackling and another voice from the *Esperanza*. "It's Paul."

Watson smiled broadly. "Is that you, Ruzycki?"

"It is."

"I was wondering when you'd say hi." Paul Ruzycki was first mate of the *Esperanza* and an old friend of Watson's.

"Glad to see you guys," the disembodied voice said. "Merry Christmas. Wishing you and the crew the best."

"Was that you waving the Canadian flag?"

"No, I was waving out a porthole."

Watson laughed. "Hey, we've got some Christmas greetings for the crew we'll be sending over later."

"Thanks. Keep up with us and we'll see you shortly."

Watson cradled the mic. Later that afternoon he gathered Christmas messages from the crew to e-mail over to the Greenpeace ships. I think for the captain they were less about holiday good cheer than an opportunity to further prick the eco giant. He made sure he led with Emily's earnest quotation:

> What we have in common with the men and women in the Greenpeace ships is that all of us are down here in these remote and cold southern waters to defend and protect some of the most intelligent and sociable mammals in the world—the gentle and great whales.
>
> —Emily Hunter, daughter of the first Greenpeace president and cofounder, the late Robert Hunter. Quartermaster on the *Farley Mowat*.

He attributed his own message to: "Captain Paul Watson—Greenpeace cofounder and former Greenpeace director 1972–1977." In fact, Greenpeace did mention Sea Shepherd on its website: a single line in a crew member's blog about the *Farley* appearing in the area on Christmas day, along with a link to Sea Shepherd's website. But that was it. The website editor had probably been reprimanded.

The pressure dropped. The storm raged. By noon the barometer had plummeted to 28.80. Everybody—the *Esperanza*, the *Farley*, the *Nisshin Maru*—went with the gale because it was the only place to go. West-northwest. There was no getting off the ride. By 0900 Watson had lost the other ships on the radar. At 1100, the *Farley* skewed and caught the wrong looming forty-footer and rolled harder than she had yet. The port bulkheads for a moment became the deck—in laymen's terms, the wall became the floor. There was an ominous thud from up forward. Big Chris Price came into the green room looking as though he was about to cry. The flying inflatable boat had snapped its tiedowns, busted loose from its cradle, and cracked itself against a stack of strapped-down steel. It was a broken bird.

The ship was weirdly quiet. Quieter even than after the last letdown. The pall had a different quality. Mandy, the older animal rehabilitator and kayak guide, said that just before the near-impact the green room was a scene. Some were running around yelling that we were all going to die. Others were sitting cross-legged on the deck staring into space as if they suddenly got it—understood that dying was definitely one likely option.

Just before lunch the Associated Press called Watson for an interview, which he gave with relish on the sat phone. By that night the story was picked up by *Newsday*, the *Washington Post*, the *Mercury News*, China Broadcast.

Watson issued another press release: "Sea Shepherd Requests the Australian Navy to Keep Peace in Antarctica." I laughed. If he had his way he'd have every warship in the hemisphere down here trying to board or fire on the *Farley*.

The barometer dropped. The old *Farley* strained and groaned. Water froze to outside steel and gave the ship a menacing varnish. She labored downwind as the black walls of erupting water hurtled from the southeast and surged beneath her. Watson didn't want to keep trending north. He was being taken farther and farther from anyplace he was likely to intercept the whaling fleet again, which was down near the ice edge.

When he came on watch after dinner he ordered the course changed to southeast, directly into the seas. The bruised and exhausted *Farley* was asked to about-face. Her speed was cut in half. The pounding was immediate. Her bow reared skyward and she launched airborne off the first wave, crashing down with a terrific shuddering and green water over the bow. Wave after wave like that. They were not diminishing, they were getting bigger. The storm was still building. Some of the waves were easily four or five stories.

Chris Aultman rushed onto the bridge half an hour later. He was dripping wet. He said the rotors of the helicopter couldn't take the stress of the head-on wind and asked the captain to turn the ship around again. Again, Watson did. Before I left the chart room I checked the barometer there against the one on my watch: 28.75. Still dropping. As I gripped the polished wooden handrails and swung down the three flights of stairs against the swinging of the ship, trying not to go airborne, I saw no one. Battened down. Eerie.

That night in my bunk I lay in the dark and thought about the trade the Sea Shepherd Society had just invited the Japanese to make: all hands lost and the end of Japanese whaling. I knew that to much of the world, Watson and many of his crew would be deemed insane. Maybe he was. Certainly, from the bridge of the *Nisshin Maru*, watching the much smaller ship hold its course to the brink of destruction must have sent a chill of cold horror through the officers. Honor was one thing, murder and suicide on the high seas another. But I also thought about the whales, swimming tonight in their herds and pods through the islands of ice, families and groups of families, in numbers that for many species were 1 percent to 3 percent of what they had been 200 years ago.

I did not think he was exactly insane. Countries around the world pledged to protect the whales and codified that promise in treaties and laws, and yet the protections meant nothing. They were all on paper. In reality the whales of the Southern Ocean, of all the oceans, were as vulnerable as if there had been no treaties at all. Japan's fleet allotted itself whatever number it wished to kill, endangered and non-endangered species alike, and came down and took them. It shot them right over Greenpeace's head. The whales could not advocate for themselves. They had no allies on the entire planet who were willing to intervene at all costs, even their own death—except Watson and Sea Shepherd. What was insane about that? Human beings were willing to lay down their lives for territory, resources, national honor, religion. What was more insane about being willing to lay down your life for another species? Whatever one said about Watson's methods, they were relatively effective. His campaigns against the English, Irish, and Scottish seal hunts in the early 1980s shut them down for good. His battle against the Canadian seal hunt had halted the slaughter in 1984—for ten years.

■ ■ ■

When I woke up on December 26, snow was blowing afresh and we were still in the teeth of the gale. Soon after, Watson got sick of being pushed west and north where he didn't want to be, and wheeled the *Farley* around again, this time to the southeast, 140 degrees. Quartering up into the waves. If he had to, he'd tack against and away from the waves, southward, ever southward, then move west along the ice edge. He wasn't going to get pushed around any more. He had a hunch that the Japanese would resume whaling in Porpoise Bay—whose mouth was 270 miles to the south of us—and his hunches, especially about finding his adversaries, had been uncannily good over the years.

Aultman had been out early, tending to his exposed helicopter. He'd fashioned some wood props for the rotors, with wood seats for the blades, to hold them stiff and high in the relentless wind. He was mad at himself for not having done it sooner. He envied the *Esperanza*'s helicopter hangar. The chopper was not meant to sit out, to be rattled by wind pressures that approached those of flight but came from erratic directions while the rotors stood still. It was not meant to be washed by salt spray for days at a time. It was like leaving your lawnmower on the beach for a week, at the edge of big surf in an onshore wind. With the chopper exposed on the heli deck as it was, having been shaken to its bones in the storm and washed with seawater, having undergone untold stresses in the feints back to the east, with the rotors bending against their tethers nearly to the deck—given all that, all he wanted to do was fly.

Aultman gave Watson the nod, and the captain turned the ship back up into the waves, heading southeast, taking the hammering seas on the port bow. The pounding was immediate. The *Farley* launched over every foaming thirty-foot mountain. Because we were not head-on, she rolled too. Some rolls were a lot harder than others. At 1110, when Watson made the turn, whoever had managed to be sleeping would have woken up with a jar. The ship was still ghostly, seemingly devoid of people. Half seemed to be seasick and the other half suffering from some sort of flu. Even stout Marc, who'd had a lot of time at sea in rough weather, was nauseated. Kristian came up long enough to tell me that ever since the encounter with the *Nisshin Maru*, he had been overtaken with exhaustion and nausea and just wanted to hibernate.

I went out into the whistling relative shelter of the starboard bridge wing and held tight to the rope rail and watched the snow and spray fly by. Not a single fulmar or albatross or petrel veered or dived around the ship. The whitecaps churned off the tops like snowy banners and were shredded instantly.

I loved this. The raw beauty of storm. I'm not sure why, but our close encounter had not scared me. I had spent half my life kayaking whitewater rivers, and the power of the ocean whipped to such a frenzy filled me not with fear but with a profound and grateful awe. I did not want to die out here either, but I could not, honestly, think of a better place to be.

When Hammarstedt came onto the bridge at noon with his watch, he looked at the wood blocks in the bearing slot and said to Lincoln, "This was Paul's idea?" Watson had gone below for a few minutes.

"Yes."

"We were getting too far west?"

"Way too far. He really wants to get south to Porpoise Bay."

Hammarstedt glanced at the GPS. Five knots. From over nine heading with the weather.

"We'll go south for a while," Watson said when he returned. "We can't really go directly south without getting the shit kicked out of us. So it's either northwest or southeast as far as I can tell. If they are going south—once we get to the ice edge, then we can move west."

The *Farley* quartered bravely into the gale. The watches on the bridge changed quietly, with Mia clinging to the radar hood and watching the screen, and wishing to God that her stomach would not revolt again; with Emily clutching her water bottle, and responding to orders from the officer with fluttering eyelids and an unconscious touch of the

Dramamine patch behind her ear. The ashes of her father—the Robert Hunter who started Greenpeace and committed his life to trying to get humans to care for the planet—were under her bunk in a box, waiting to be scattered in one of the last truly wild places. The J. Crew crawled out of their berths only to watch a DVD of the TV show *Lost*. Even forest wildman Gedden didn't look well. He had just tacked up a self-published calendar on the corkboard in the mess called "Earthfirst! Hotties," which featured a full-page photo of Inde—naked behind strategically-placed foliage—Miss February, I think—arching back to shoot an arrow into the canopy. Gedden himself, Mr. January, Caliban of the Cascades—stood fully frontal holding a giant monkey wrench. Other nudes stretched demurely on their tree-sit platforms, which didn't look big enough to play cribbage on. Hard to imagine feisty Inde up there for six weeks at a stretch. The calendar was popular, and despite the collective nausea of the crew, was already getting dog-eared.

Gedden sat on the bridge on his watch with a sullen determination not to get seasick. The pounding and the canted launching of the bow over each wave were exhausting. Gedden said at one point, "All meat eaters should be shot."

In the radio room, Watson was playing solitaire on the green virtual felt of the laptop on his left, while he sat at the main keyboard and pounded out more correspondence and invective. He had international media on the hook with the dramatic Christmas encounter, and he was going to play that fish for all it was worth.

"There's a rumor," he said with glee as I leaned into the room, "that the Japanese are sending down a warship. I have it from a reliable source in Japan."

Good God. He wouldn't make that up—or would he?

"Are you winning?"

"Huh?—oh, that. You know Al Johnson, one of the first Greenpeacers and a great supporter of Sea Shepherd, once said if you're ever lost, really lost in the wilderness, just start playing solitaire."

"Why?"

"Start playing, and before you know it someone will be looking over your shoulder telling you how to play the next card." He laughed. "What time is it?"

"Seven-fifty."

"Guess I'd better get up."

On the bridge, I asked Watson for an update on his plan.

"The way I figure it, they've got to come into the ice again. Once we get down here to the ice edge, we can send out the helicopter." He watched the bow gash into the trough and take a wave over the rounded whaleback. "Doesn't make any sense why they'd be out here—the storm isn't even down in the ice edge. Maybe they thought they'd lose us in the storm. Our mole won't tell me where they are. Sometimes he does, sometimes he doesn't."

All day the waves and wind seemed to be backing off a little. Waves that towered high over the bridge as they thundered in were more rare. They were just big enough now to obscure our view of the horizon from the bridge—in the twenty- to thirty-foot range. The *Farley* hit them with her port bow, rolled way over to starboard as she met each one, and then she launched, undaunted over the top, as if she were trying to leave the sea altogether and follow the fulmars and petrels to wherever they had gone that wasn't here. But her 800 tons were made for nothing but water and, as became more and more apparent, storm. As she came over and dived into the trough, she righted and settled and nodded to the next—all with a long, unhurried rhythm; a staunch, deep-bellied near-ease that softened the transitions, like a fat lady dancing.

Watson sat in his captain's chair, tapping away on a book he was writing about religion called *God's Monkey House*.

"You know," he said. "Steller's sea cows went extinct twenty-six years after they were discovered. They were the size of elephants. In the Aleutian Islands. They didn't have any enemies, so they weren't afraid of humans or anything. They just lay there. The Russian fur traders killed them all off. If they wounded one the others would come and try to comfort it. Really gentle creatures. I never understood how anybody could do something like that."

The few of us who were on the bridge went silent.

Big Marc rubbed the blond stubble beneath his lamb-chop side-burns and watched the *Farley*'s bow take another hit. He blinked and his eyes were wet. "What we do to the ones that have no defense," he said. "Yah, well." He got up and went below.

"Do you think we're going to hell for what we've done?" Lincoln said.

"Hell?" Watson guffawed. "Heaven and hell were defined by two poets, Dante and Milton. None of this is in the Bible. Sort of like 2,000 years from now we believe in the Force and the Jedi Knights."

He tapped away on his laptop. "Right now I'm writing about the Christian pantheon of angels. Did you know the wingspan of a cherub was fifteen feet? Ten cubits. References to angels in the Bible: 177 in the Old Testament and 179 in the New Testament. Paul comes from 'Paulel,' which is Paulus, small; and el, angel. The 'angel of small things,' which I think is appropriate.

"Here's another thing: an unnamed angel of peace and an unnamed angel of truth opposed God's creation of man—so God burned them."

I turned to glance out the front window just in time to see a huge black wave. It must have been forty feet. The *Farley* dipped dutifully into it, and the wave crashed on the bow and swept the main deck, and slammed into the windows and over the monkey deck above.

The storm was renewing itself with a vengeance. Just when I thought it was easing off, the wind tempered itself to a dangerous hardness, slugging the bridge with what seemed more and more like solid blows. And the waves steepened and came faster, breaking more often over the bow. They were also bigger. We were heading generally south, as due south as the *Farley* could withstand; and where we were going there would be more and more ice in the water. I did not want to think about what would happen if one of those steep waves broke over the bow with a five-ton wedge of glacier ice in the middle of it— what would happen to the tired old patched hull.

Stout Geert was still at the chart table, illustrating the logbook with a gorgeous full-page picture of an iceberg with a blue grotto.

"Everybody's sick," I said. "Seasick or sick with a cold or flu."

Geert's eyes twinkled as he picked out a gold pencil from his box. "Or homesick, or sick with something."

"Or lovesick. For Sandra Bullock."

Geert pretended that he no longer understood English and began to trace a figure in the ice cave.

I said, "I think they have post-traumatic stress syndrome. I think everyone realized they could have really died. Heading straight into thirty-foot waves choked with ice probably doesn't help."

I looked at the logbook. Geert was drawing the Virgin Mary. She was in the ice cave, extending one hand toward the sky, lit by a golden yellow beam that angled down from the heavens. Dutch artists just could not resist that beam. You could walk on it if you had to.

Holy Mother of Ice. Immaculate Virgin of the Dibble Glacier.

"Geert."

"Huh?" He was bent over his work so that his bushy beard brushed the chart table.

"What is that?"

"That's for Paul. He's gonna hate it."

At 2200 Watson got a new fleet position from his source on the *Esperanza*. Fearing discovery, the source had called a brother in North America who had just called Watson on the sat phone.

The *Nisshin Maru* was at 63°30'S, 137°05'E. That was almost 200 miles ESE of us, and not far from the south magnetic pole. The report said the factory ship was heading fifty-five degrees, or nearly northeast, but was generally going in circles.

Watson didn't understand it. "Maybe they're waiting for the killer ship to come back from Hobart," he said. He would hold his course to the southeast, aiming for the east side of the long cold protrusion of the Dibble Ice Tongue. Then he would work along the ice edge to the west, where he thought they would have to return.

16

The Law of the Sea

On December 27 the ship was even more quiet. Alex, who must have slept only a couple of hours, came onto the bridge muttering, "This is insane; where is everybody? What is this, a party cruise?"

The first mate was disgusted. "I'm tempted to ring the alarm on the console," he said. "Even Colin isn't responding. Everybody is in their berths."

Watson barely noticed. He was talking about his only child, his daughter Lani. He said, "When Lani was in sixth grade the teacher asked the class what was the function of government. Lani said, 'The government is a bunch of people who get together and plan how to kill people and animals.' That raised a few eyebrows. Later, her teacher asked if anyone knew why they called it the horse latitudes. Lani said because when the ships were becalmed they threw horses overboard to tow the ships. The teacher said, 'That sounds a little far-fetched. Where did you learn that?' Lani said, 'My father's a sea captain and you're just a teacher!' She got in trouble."

Watson said the horses would tow the ships for miles. Once the ships got some wind, the crew would cut the horses loose and abandon them. "One of the horses' caretakers once killed a captain and first mate over abandoning horses."

He got out of his chair and went to the radio room to scan his morning's e-mail. He came back onto the bridge shortly, flourishing an open letter from the director general of Japan's Institute of Cetacean Research, Hiroshi Hatanaka, to Junichi Sato, the campaigns director of Greenpeace Japan:

December 27, 2005
Mr. Junichi Sato
Campaign Director, Greenpeace Japan

Sir:

. . . The captains of the *Nisshin-Maru* research fleet, which is registered in and operated under the laws of Japan, have the duty of ensuring the safety of our crew and vessels. . . . Any escalation of Greenpeace's violent activities would correspond to piracy as defined by Article 101 of the United Nations Law of the Sea.

. . . Your website shows that Greenpeace is now cooperating closely with the Sea Shepherd Conservation Society to further your activities in the Antarctic.

The Sea Shepherd boat, the *Farley Mowat*, has already foolhardily tried to approach the *Nisshin-Maru* and deployed a mooring line with the intent of entwining her propeller. These are extremely dangerous actions.

It is widely known that Sea Shepherd has engaged in criminal and violent activity in the past, such as setting fire to and sinking whaling vessels in Iceland and Norway and fishing vessels in Spain and other countries.

Sea Shepherd is a terrorist organization—their members use threats of violence, sabotage, and an open disregard for human life in furthering their cause. From your recent activity and acknowledgment on your websites, I can only conclude that Greenpeace is colluding and cooperating with Sea Shepherd; that you have recognized Sea Shepherd and are proclaiming to the international community that Greenpeace itself is kindred to the same violent and ecoterrorist approaches that Sea Shepherd is well-known to undertake.

Hiroshi Hatanaka
Director-General, Institute of Cetacean Research

No better way to mortify Greenpeace than to accuse it of collaborating with Sea Shepherd. The moral high ground its officials had occupied in their scornful rejection of Watson's pleas for cooperation would be snatched out from under them. They would have no choice but to attempt to further distance themselves from Sea Shepherd. Divide and conquer. By accusing Greenpeace and Sea Shepherd of piracy and terrorism, the letter also set the legal stage for severe countermeasures. Under the law of the sea, acts of piracy constitute grounds for arrest, seizure of the perpetrating ship, and even deadly retaliation in self-defense. It was a warning shot across the bows: we are defining your activities as piracy, and we will respond in accordance with international law.

Greenpeace responded immediately. A lead story in the *New Zealand Herald* the next day said:

Pia Mancia of Greenpeace New Zealand denied the group was linked with Sea Shepherd and said there was no mention of it on Greanpeace's website.

"We are two separate organizations—we wish them well, but we do not work with them. They are going down there, we can't stop them, but we hope they will work as safely as us."

. . . In a Christmas greeting to Greenpeace, Sea Shepherd crewmember Pawel Achtel of Australia quipped, "We are not violent! We are collecting Japanese ship samples for scientific research," an apparent reference to Japan's contention that it only kills whales in Antarctica for scientific research.

Things were heating up on the diplomatic and media fronts, which is exactly what Watson was trying to do. He may not have been able to bring the bow of the rugged *Farley* into contact with the hull of the *Nisshin Maru*, but he had scared the Japanese and ignited this flurry of international accusations. No wonder he looked so happy, humming away on the bridge, while his ship got pummeled by icy seas and his traumatized crew hibernated.

An article the next day on the Cybercast News Service, CNSNews.com, was headlined "Green Activists Target Japanese Whalers, Bicker Over Tactics":

> . . . Watson this week claimed that the main Japanese ship tried to ram the Sea Shepherd vessel and said only the release of mooring line from his boat's stern prompted the Japanese skipper to alter course so as to avoid having his propeller fouled.
>
> Several days earlier, Watson issued a statement accusing his former Greenpeace associates of obstructing Sea Shepherd's efforts by not telling his group the location of the Japanese fleet, which Greenpeace had managed to locate first. . . .
>
> Watson added, however, that his group and Greenpeace were "both pursuing the same target. Greenpeace can deny we are allies all they wish, but the enemy of my enemy is my ally, like it or not," he said.
>
> Greenpeace was, indeed, trying to distance itself from Sea Shepherd, especially in light of a letter released by the head of the Japanese whaling body.

The article closed with this kicker:

> Last week, the New Zealand government released what it called a "damning scientific critique" of Japan's whaling program, saying that its stated research objectives were scientifically unsound and those few objectives that were relevant could be achieved by nonlethal means.

Watson didn't take long to fire back with a response of his own to Hatanaka's letter. Hatanaka had called him a terrorist, in a public venue. Hatanaka's position was publicly funded. It was not a stretch to say that the government of Japan was openly accusing Watson of terrorism. This was a serious accusation, all the more so given the current international climate, and especially in light of what the French Special Forces did to the Greenpeace ship *Rainbow Warrior*. Once a government had someone in its sights as a terrorist, the results were usually not pretty.

Watson blasted:

> It is Japan that is violating international conservation law. Specifically this can be summarized as:
>
> 1. The Japanese are whaling in violation of the International Whaling Commission's global moratorium on commercial whaling. The IWC scientific committee does not recognize this bogus research that the Japanese are using as an excuse.
> 2. The Japanese are killing whales in the Southern Whale Sanctuary.
> 3. The Japanese are killing whales unlawfully in the Australian Antarctic Territory.
> 4. The Japanese are targeting fin whales this year and humpback whales next year. These are endangered species and thus this is a violation of CITES, the Convention on the International Trade in Endangered Species of Flora and Fauna.
> 5. The Japanese are in violation of IWC regulation 19(a). *The IWC regulations in the Schedule to the Convention forbid the use of factory ships to process any protected stock:* 19.(a) It is forbidden to use a factory ship or a land station for the purpose of treating any whales which are classified as Protection Stocks in paragraph 10. Paragraph 10(c) provides a definition of Protection Stocks and states that Protection Stocks are listed in the Tables of the Schedule. Table 1 lists all the baleen whales, including minke, fin,

and humpback whales and states that all of them are Protection Stocks.

6. *In addition the IWC regulations specifically ban the use of factory ships to process any whales except minke whales:* Paragraph 10(d) provides: (d) Notwithstanding the other provisions of paragraph 10 there shall be a moratorium on the taking, killing, or treating of whales, except minke whales, by factory ships or whale catchers attached to factory ships. This moratorium applies to sperm whales, killer whales, and baleen whales, except minke whales.

What law has Greenpeace or Sea Shepherd broken?

It is not illegal to interfere on the high seas against their illegal whaling activities. In fact we are legally authorized to do so in accordance with the UN World Charter for Nature.

The United Nations World Charter for Nature states in Section 21:

States and, to the extent they are able, other public authorities, international organizations, individuals, groups and corporations shall:

(c) implement the applicable international legal provisions for the conservation of nature, and the protection of the environment;

(d) ensure that activities within their jurisdiction, or control do not cause damage to the natural systems located within other States or in the areas beyond the limits of national jurisdiction,

(e) safeguard and conserve nature in areas beyond national jurisdiction.

And finally Section 24 states:

Each person has a duty to act in accordance with the provisions of the present Charter; acting individually, in association with others or through participation in the political process, each person shall strive to ensure that the objectives and requirements of the present charter are met.

The weather changed. I woke on the morning of December 28 to the easy gurgle of water along the hull at my head, and a steady, easy pitch; the bump and crunch of an occasional growler colliding and rolling off. From the deck it was a different world: an innocent, seven-foot swell, light fog, fulmars and petrels circling the ship. The rhythmic thresh of the bow wave. The waves had shifted south a little to the ESE and we were running straight into them.

On my way back down the hall, I saw Mandy corner Electrician Dave outside the bosun's locker.

"You're a hunter, right?"

Dave looked up at her. Mandy is formidable: tall, broad-shouldered, with a hawk's face—beak of a nose and dark, close-set eyes. She towered over the little Australian. Dave knew she had spent a considerable amount of time doing "hunt sabs"—risking her life trying to sabotage hunters in California and Colorado.

Dave cocked his head. "Aw, yeah. Well—the only thing I hunt is introduced nonnative feral species." Wow, I didn't know he had it in him; I'd never heard him so eloquent. Dave held up two fingers. "Foxes and pigs. They're doing terrible damage. Terrible." He shook his head morosely. "Nobody's doing anything about it, you know, so we do."

He looked up at his adversary, measuring the effect, the way you'd test a wire with a voltmeter, and he grinned, showing his missing tooth. As I passed, I thought he was smiling because he knew he'd escaped this time, barely.

By almost noon we were at 64°45'S, forty-eight miles from the Dibble Glacier Tongue, and just to the east of it. Almost to the coast. The fog was thickening, but the mood of the crew was getting better all the time. At lunch, Chris Price told a story about bringing his mother-in-law into a whorehouse in Nevada to use the restroom; he must have been feeling better about the wreck of his FIB Thing.

By four in the afternoon we were sailing slowly into a luminous, fog-shrouded snowfall. The seas were maybe eight feet. The dark water was peppered with growlers, and Alex's watch steered through them carefully, in concentrated silence. We were forty miles south of the south magnetic pole, and though the GPS showed us going southeast, the magnetic compass in the binnacle was skewed to the northwest, 180 degrees wrong. Amazing that anybody could have navigated down here before satellites. We had passed sixty-five degrees south and should come to the ice edge soon. We were 100 miles from the whaling fleet's last communicated position.

Watson came onto the bridge and told Alex to slow and just head into the waves.

"We'll wait here for a while. I haven't heard anything from the Greenpeace ship for two days. They were probably given new orders."

By nine, the *Farley* was pushing slowly into a dense fog. The bent, and rusted thermometer on the bridge wing was still pegged at thirty-two degrees, and the fog verged—half a degree colder and it would shiver into a veil of ice. Visibility was under 100 yards. Mia was steering and Lincoln was alternately looking ahead and peering at the main radar screen.

"Object on the radar," he said; "17.4-mile range going 6.1 knots. Looks like a clump of three objects."

"Should I change course?" asked Mia.

"No, just watch it," Watson said. "If they're harpoon vessels they'll just move off." He looked over his shoulder at Aultman. "We need to get the helicopter up," he said.

Aultman blinked at him. "Yeah, I want to. I can hover over the ship." He laughed.

Watson got up and went to the radio room. He came back in a few minutes. "Got an e-mail from the *Esperanza* dated about one hour ago. And he says they're doing nothing. The *Nisshin Maru*. Up, down, and going nowhere. The harpoon boats, they've got to come back. They can't go whaling with a whale tied to the side." He glanced again at Aultman. "If he's a real helicopter pilot he'll go up in this!" he said. They were within 100 miles; the harpoon boats were probably closer.

He wanted one clear swipe. He had said yesterday that you could almost completely crush the bow of the *Farley* and she would still float.

The three objects on the radar had stopped moving and looked more and more like icebergs. I went to bed.

A rapping on my cabin door woke me up from a sound sleep. I sat up in the dark. The engine was throbbing low, and the plashes and gurgles were almost languid. The *Farley* must have been idling at a near standstill.

"Yeah?" I hit the light button on my watch: 0330.

"Come up to the bridge: we've made a contact." It was Colin, the SWAT-clad butcher.

I slipped into fleece pants and a jacket, grabbed the dry bag and dry suit, and bounded up the three flights to the bridge. It was crowded up there. I pushed through to my usual spot by the forward port window of the bridge.

Outside was a twilight of drifting fog and slick, almost windless water. The fog off our port bow held the shape of a ship, maybe a mile off. It was not large. I looked down at the main deck and just aft of the whaleback five deck hands were working fast, coiling a green mass of rope and stashing it forward in the forepeak.

"He's got his port beam to us," Trevor said.

"Wiping out the damn Patagonia toothfish. I don't care if it's legal or not."

"That's a beautiful ship," Trevor said.

Apparently just after midnight the *Farley* had been cruising westward toward the ice edge projecting from the Dibble Glacier when the watch had encountered several bright pink longline buoys. They snagged them with the boat hook. One had a radio transmitter. When they tried to pull the buoys in, they saw that they were attached to a longline. They hauled it up and ran it through the big winch on the bow. They rolled up almost a mile and then started picking up the baited hooks. Big two-inch hooks on leaders with chopped bait. Must have been pretty fresh as there were no fish on the line. The line had been set on the bottom at almost 4,000 feet. It took three hours to retrieve it all. Supple, expensive, ¾-inch rope, about $10,000 worth.

Marine channel 16 crackled. Spanish. Gunter, our German Brazilian, picked it up.

"*Sí?*"

"*Vamos, arriba—diez y siete.*"

Watson flipped the channel up to 17.

Gunter talked rapidly. An amiable enough voice replied on the other end. "OK, OK, *entiendo*—unh-huh, *americano. Qué barco?*"

"*El Farley Mowat*—what are you doing?"

A pause, just long enough to register as mild shock. It was improbable enough to meet another boat down close to the ice edge off the Adélie Coast, west of Commonwealth Bay. But then to have it demand what you were doing there. The voice answered in a staccato burst. I caught the word *Uruguayeno*. And *pescando*. No kidding.

Gunter turned to us. "Are fishermen. Fishing for cod."

"Like hell," Watson said. "Nobody fishes for cod down here."

"We have a license—*vamos*," the voice added, not without some anxiety.

Gunter said to Watson, "They are pulling the lines here. They are asking is it OK?"

"Is it OK?" Watson repeated. His face hardened into the all-business, almost blank, predatory look I had seen before. What the Uruguayans were doing was fishing for Patagonian toothfish, aka Chilean sea bass, an endangered species. The fish was on the verge of being wiped off the face of the earth. One trip to Antarctica could net these men nearly $1 million. For that, they were willing to send a very ancient creature into oblivion. What Watson wanted to do more than anything else was rev the *Farley*'s engines and make a beeline for the boat. He wanted to ram it. He wanted to crush its line spool and winch, crumple its stern. He'd done it before—a few nights ago I'd seen the film footage of an attack by Sea Shepherd on Japanese drift-netters—the surprise and terror on the faces of the fishing crew, the bruising impact, camera frame jarred and blurred, then crushed steel, the power blocks crumpled and skewed.

"Ask if they have a license from Australia," he said to Gunter, not taking his eyes off the ship, which shape-shifted in the mist.

"Ah, no," replied the voice, more obviously nervous now. "*Uruguay, uruguayeno.* We have a license from Uruguay. We have a license for here and for the Ross Sea."

Watson considered no more than a second. The answers indicated that they were most probably poachers. He couldn't be sure.

"Ask if they have seen any Japanese whalers," he ordered Gunter.

"Ah, no, no." the voice sounded relieved. "OK, OK. Good luck. Hope you find them."

Wessel came onto the bridge. "There's about three miles of longline in the forward peak," he said. All the line they had confiscated and stuffed out of sight.

Watson nodded. Gunter and the fisherman were talking rapidly. Gunter turned to Watson. "The number of registry is 5151. I asked the number of the license. He said he didn't know the number of the license, the captain is asleep. The *Paloma* is the name of the ship. Captain is Pérez." *Paloma*, the Dove—beautiful name.

"If it's fishing here, then it's going after Patagonia toothfish," Watson said. "I don't want to piss off the Australians by harassing someone that's legal. There are only a couple of licensed boats. Last thing I expected to see down here—a longline." He worked his jaw side to side and rubbed his beard with the back of his hand.

"This species is on the verge of being wiped out. Here the concern is for the fish itself. Tangle with this guy and he's legal and we'll really get in trouble."

Watson looked thoughtful. He looked at the crew on the bridge and smiled. "Years ago the Japanese were putting out a fifty-mile drift net. They were putting it out and we were reeling it in."

It was 0400. The watches changed. Hammarstedt and his quartermasters, Emily and Darren, handed over the bridge to Alex, Gedden, and Gunter. Everybody was already there anyway. Watson stayed in his captain's chair.

"OK," he said to Alex. "Keep going west, 270 degrees. We'll have Tim back at the office check up on the registry." Watson had decided to keep his focus on the whalers. He got out of his chair. "M— contacted me at midnight. M— says they're not making any sense. Going this way, that way, not hunting. No sign of the killer ships. He gave me the position at midnight. He had a friend call from Toronto again. The friend said he can't call directly or e-mail. They're starting to suspect. He said to keep a position to the south. Now they're directly north of Porpoise Bay, which is where I thought they were going."

The *Nisshin Maru*'s last position was: S 63° 37', E 131° 33', and heading WNW, 280 degrees. We were now at S 65° 19', E 138° 49'. That put the whalers about 200 miles north of Porpoise Bay and about 200 miles to the west-northwest of us. So Watson, it seemed, had a good instinct in wanting to follow the ice to Porpoise Bay. If we kept steadily cruising west, and the weather held out, we could be there in twenty-four hours and wait.

Mid-morning, the ice continent was less than 100 yards off to port— an undulating false coast of broken, floating ice that stretched south to the real shoreline. It looked like land, but when the swells came through all of it rose and fell gently.

Other than the monoliths and an occasional fat seal dozing away the summer morning, everything was unrelieved grayish-white as far as one could see. The dark blue water of the Southern Ocean met the ice edge in a sharply defined, wavy line. We followed the seam westward. So did squads of Adélies, which porpoised fast just off the pack, and passed us.

At 1040 it began to snow. A thick fall of dry flakes that swirled over the main deck in almost no wind, and the fog thickened with it. We slowed and followed a smooth, two-foot swell, paralleling the ice edge, very close, perhaps thirty yards off, like a blind man holding on to a rail. Groups of two and three penguins stood and watched us pass, and a large leopard seal lifted his broad head and then put it back on his ice pillow.

Across the skirt of ice, about thirty miles to the south, were Victor Bay and Porquoi Pas Point. Why Not indeed? I laughed at the juxtaposition. Was it two different explorers who named them? Or one moody Frenchman, who had entered the bay triumphant, and then succumbed to existential post-victory ennui as he sailed out to the headland?

Watson sat in his captain's chair on the bridge and watched the peacefully falling snow. He was not feeling nice. He was gnawing over the Uruguayan poachers, whom he wished he could have rammed and sent limping home.

"I don't understand how a species that's been listed as threatened can still be fished. Remember the orange roughy? Don't see that anymore. All gone. Takes the fish forty-five years to mature to breeding—forty-five years to get it on. We treat all fish the same. Salmon live something like four years, and sharks are theoretically immortal. They don't die of old age—that's why they only produce one offspring. So when you take 100 million out of the ocean every year, it has an incredible impact. 'Perfect killing machine!' Most sharks don't even kill, they're scavengers."

Trevor appeared on the bridge. The captain had asked him how much fuel we had left and he'd been sounding the *Farley*'s seven tanks. Watson didn't want to get so far west along the coast of Antarctica that he would pass the point of no return and be unable to get back to Australia. There was talk of perhaps abandoning that idea altogether if the whalers continued west of Porpoise Bay. In that case he might try instead for Cape Town, South Africa.

"Seventy-eight thousand-plus liters, which translates into 66.9 tons, which translates into twenty-eight days. We're not going to South Africa. We're going back to Australia." He twisted his mouth into a wry smile. "Or Antarctica. It just means we have to stay in this area. South Africa is 5,000 nautical miles farther."

"What's the water temperature?" I asked.

"By touching the intake pipe I can tell you it's fucking cold. The sea temperature should be about two degrees centigrade."

"In other words," Lincoln said, "you don't have any fucking clue."

"I know a lot more than you do," Trevor snapped. He went to the door and turned. "Any more questions?" He left.

For the rest of the morning we churned west. Watson wanted badly to send up the chopper to look for killer ships he figured must be down here somewhere along the ice edge, probably just ahead to the west.

"I wonder if it unnerved the Japanese, coming out of the mist like that?" Watson mused as he looked out into the shroud that kept the helicopter decked for now. "All we need is one more whack at them."

Mia steered now, and Lincoln looked ahead with binoculars for growlers. Down on the main deck, just under the forecastle, Alex, Kalifi, and Inde were still untangling a heap of the thick green long-line rope. Two Cape pigeons circled over them, and three birds I hadn't seen before. They were small and black, with short wings and white rumps. They looked more like shore birds.

We were cruising at a steady nine knots, the fastest we'd gone in days. Aultman came through the port bridge wing door.

"Thirty-four degrees. Woo-hoo! Heat wave! I want to give her a bath."

"Don't you wish you had a mechanic to check her over after all that storm?"

"Tell you the truth, I'd be happy to have someone who'd even seen a helicopter before." The former marine continued on through, going to look for Casson or Dennis to help wash the chopper.

If it didn't feel exactly like a heat wave, it felt like a reprieve. The *Farley* seemed to loll on the little swell, and the sea was smooth. There were more birds now than there had been at any time before, and the penguins zoomed by us in little families. Except for a line of dark snow squalls to the north, the sea, for the first time I could remember, seemed devoid of menace. In the bosun's locker, Geert hummed while he used the jigsaw to cut a sea turtle out of scrap plywood. He was beginning to populate the wainscoting of the bridge with these cameos of wildlife, all painted with swirling tribal designs. He took the job of ship's artist very seriously, and put in overtime every day. I noticed a dab of cerulean blue decorating the end of his beard. The rest of the crew was recovering and back at work.

Kailifi called a meeting of the deck hands and sat in the bosun's locker in a folding chair with her legs crossed, sequin sandals winking, sunglasses propped fashionably in her unruly blond hair—though there hadn't been any sun for two weeks. The oil drums had shifted in their straps. The FIB wing had been lashed to the starboard rail and needed to be broken down and stored. There were still piles of long-line on the main deck, which Inde would supervise in coiling. A half mile of it would be rerolled onto the spool on the stern to replace the mooring line. And the old ship would get a scrub-down, stem to stern.

The wavering border of ice swung up to the northwest and the *Farley* followed. We were about sixty miles due north of the Dibble Ice Tongue, skirting the ice it shed as it broke up. Evening settled with a dimming, leaden overcast and a vanishing of fog and snow. It was the opening Watson had wanted. He told Alex to steer the ship straight into the broken field ice. The *Farley* pushed a quarter-mile in, opening a black curving swath behind her. Alex shut down the engines, and she ground to a halt. A leopard seal and a crabeater basking off to port barely noticed. They were separated by fifty yards.

Alex cut the engines, and a profound quiet overtook the ship. Only a muffled hammering from the bosun's locker, the clank of a bucket on deck, the light scrapes and shifts as the *Farley* rose and fell with the ice around her. A mile to the south of us a tall obelisk of ice, a little like the Washington Monument, caught the diffuse light, and sharpened its edges against the layered sky. The seals slept. A little while later Aultman started the chopper and did a careful half hour run-up of his engine and other systems, and then he lifted off with Kristian and took off to the northwest, heading toward the *Nisshin*'s last position. He had been gone about an hour when Watson got another communiqué from the *Esperanza*.

The fleet had broken its pattern of indirection and was running almost due west. It was still well away from the coast and was already at longitude 122 degrees east. That was about 350 miles west-northwest of us. The *Nisshin* was going fourteen to fifteen knots, about as fast as it could go. Still no sign of the killer ships. The message also said that the *Arctic Sunrise* was 100 miles behind the *Nisshin Maru* and the *Esperanza* and losing ground.

Watson called back the chopper. His target was heading farther west at speed. Even if we followed, the gap would be increasing all the time. We were getting low on fuel. If we went farther to the west we wouldn't have enough to get back to Australia, and Trevor didn't think we'd make it to the southern tip of Africa at Cape Town. In a few days we'd be past the point of no return. The seas were calm now, but if we hit more bad weather, we'd use our remaining diesel at an even faster rate.

Watson was unfazed. He patiently waited for the helicopter to return. The deck hands lifted the center Zodiac out of its cradle, opened the winged doors of the fish hold, and lowered the ungainly FIB wing back into obscurity. Marc continued to try to unbend one of the FIB's broken stainless couplings, but no one believed the Thing would ever fly.

Even in the night gray and scudding overcast, the ice that surrounded us was almost too bright to look at. It was easier looking toward the dark blue of the open ocean. The light breeze and easy swell continued out of the southeast. A dark curtain of snow squalls hung across the horizon in that direction, over ice and water. The weather down here didn't rest for a second. In minutes the storm could be on us. Which was the most dangerous threat to Aultman.

He returned to the ship at 2225, and Watson gave him the time he needed to strap the bird down, prop the blades, refuel, and oil up. At 2300 Watson called Trevor and told him to start the engines, and the *Farley* nosed a slow circle out of the ice and followed her own trail, now loosely filled in, back out to open water. We followed the ice edge north-northwest into the stubborn twilight. Watson would round this pack and continue west in chase.

Alex, who was less patient than Watson, sat cross-legged on the floor of the chart room, his back against a cabinet filled with smoke bombs, and drank from a bottle of Johnnie Walker Red. It must have been the gift from the woman in Melbourne. All Alex wanted to do was engage his enemy. The misses, the flagging chase, the danger were taking a toll on everyone. It was like a dream in which you run, but you are running through molasses. And the *Nisshin Maru*, shrouded by distance and storm, was an ominous, deadly presence that informed every decision and thought. We were homesick, apprehensive, and, more than anything the crew wanted to attack.

Alex passed the bottle around a small circle—his countrymen Geert and Marc, Gunter, Colin. It was a tough-looking crowd. If they had been seated at some bar, you might think twice about pushing through. There was another bunch in the sauna just outside—Inde and Laura, Casson, Willie. Inde had just tugged open the heavy steel door from outside and she stood steaming in the chart room. She wore overalls in lieu of a bathing suit, she was flushed, and her huge eyes shone. She'd had some wine before the sauna. She was soaked from sweat and steam, and her black hair hung straight and wet down her back. She swayed a little with the gentle roll of the ship. Watson emerged from the radio room a few feet away and said he'd just gotten word from a reliable source in Japan confirming that the Japanese government was sending down a warship to defend the fleet.

Somebody offered him the bottle, and he held up a hand—no, thanks. He was on watch for another few minutes. "It's interesting that the director general sent a note specifically mentioning Section 101 of the Law of the Sea, which deals with piracy. According to the Law of the Sea Article 105, a warship is authorized to seize a vessel that is operating in international waters as a pirate and bring the crew back to the nation for trial."

"What would be your strategy with a warship?"

"Oh, we're not going to get seized. I'm not going to recognize their authority to board us. I wasn't seized by the Russians or the Norwegians. I'll go where the whalers go. If they want to fire on us, so be it; but that would be pretty stupid. They can take the *Esperanza*." He smiled.

Inde swayed and studied him. "I bet the people on the *Esperanza* are nice," she said. "I bet some of the Japanese are nice, too."

Watson glanced at her. "Yeah, we're all fucking nice."

Inde let a smile tremble at the corner of her lips and then she turned on the captain. She was as tall as he was, and her shoulders were nearly as broad. "Hey, next time . . . When the ship was about to pounce on us . . . we were all ordered to the green room. It would have hit right there. Next time let's not sit in the green room." She shook her head. "I was thinking we were all going to perish on Christmas. I thought, 'We're all going to die.' I mean, if that would have stopped whaling that would have been OK with me, you know?"

Watson said, "Nobody was going to perish."

"Why? You didn't think we'd be sent off on the life rafts? In those waves?"

"No."

Inde swayed. "It was strange. Just seeing it—it was almost like reading in a myth or in a story—this demon coming from hell, or this three-headed dog coming out of Hades—tearing through the veil." She tipped her head. "This dark presence just there, you could feel it. Even the ship itself felt like it had to go down."

The men stared at her as if she were delivering the final soliloquy in a play.

"They must've thought we were fucking crazy," Inde said, and went below.

17

Flight

Friday, December 30, 2005

First sun in two weeks. Sun and wind. The pale blue directly overhead is so novel I go out on deck and lift my face and let the color soak like rain. The sun is intermittent, as the horizon is still ringed with clouds. Today is my day to go up in the chopper. Watson wants Aultman to take a particularly long scout out ahead of us, out to the west. We have rounded the corner of ice extending from the Dibble Glacier, and the *Farley* is now chugging due west, out in an open cobalt sea, about thirty miles off the ice edge of the Banzare Coast. The wind has shifted into the southwest, raising a choppy six-foot swell with some whitecapping.

I still don't see what Watson wants to accomplish with his recon flight, or even with this chase.

"The *Nisshin Maru* is now 400 miles away and increasing the gap," I say, pointing to the chart.

"Those guys—Greenpeace—don't know where the harpoon vessels are, they haven't seen them for days. So if we stay down to the south we may run into them. We're keeping up with the *Nisshin Maru* as best we can, and at the same time searching for the harpoon boats. Really a flanking motion: the *Arctic Sunrise* is on one side, the north, we're on the other—sweeping the corridor. Between three ships and the helicopters we're sweeping a good bit of area."

That still didn't answer my question. The heart of the fleet was 400 miles away. The harpoon vessels could easily go twenty-five knots—take one look at us, put the pedal to the metal, and be gone.

But I was glad to be going up with Aultman. We'd been on the ship a long time, and any opportunity to get off, short of swimming, was a welcome diversion. I put on my dry suit and a life jacket as well. In ocean conditions the dry suit should have been enough to float me plenty well. Aultman smiled when I climbed in.

"You're certainly not going to drown," he said over the blast of the engine.

"That's the plan."

We put on our headsets. The roar muffled; Aultman spoke through his mic. "This is pretty much the limit of conditions I can take off in. Let's pray the weather holds."

He snugged his GPS into its holder.

"If something goes wrong and we have to ditch in the water, out here we stay in the chopper. In these seas she'll bob like a cork. If she starts to turn over and we have to bail, I climb out, I throw out the life raft." He pointed his thumb to a compartment outside and behind him. I gave him the thumbs-up. Everybody knew that something had better not go wrong out here.

The *Farley* was pitching hard and Chris waited for the right timing and lifted off abruptly. Immediately we were buffeted by a twenty-five-knot headwind. He rolled away from the ship and then we tipped forward and he accelerated into the weather, which was coming now from the west. We climbed fast to 5,000 feet. The sea was a black floor rippled in long stripes by the swell, and flecked with whitecaps that sparked and disappeared in a vast and silent propagation. We angled for the blinding-white coast of the ice edge, which marched to the horizon and lay taut as a skin over the curve of the earth.

The chopper rattled along at seventy-five knots. It was exhilarating to be airborne, to be free of the waves, and to cover so much territory so fast. We saw big solitary vessels we thought might be ships, but when we flew over them they were islands of chisel-cut ice, cupping lagoons of electric blue. Nothing. No Japanese, no harpoon boats. At eighty miles out we ran into menacing dark squall lines coming in from the west and north and decided to turn around. Eighty miles. The expanse of empty ice and sea around us was almost overwhelming. We were thousands of miles from nowhere. Should we have to ditch, we were at least nine hours from a lumbering ship, a speck as small as the whitecaps. We headed straight for the ice to see what wildlife we might find and ran back along it.

The false coast was a tiling of tightly packed, geometric ice cobbles that merged in the distance ahead to solid white. The dark sea pressed it into a mimicry of coves and bays. Spires stood to the south like bone cities. Not a whale breaking the surface, or a seal to relieve the relentless mosaic.

The world below was black and white. Pure and lonely. That's what it was about Antarctica: you are either hot-blooded and hungry or you are a cold element. You are water or ice. There is no middle ground here, no compromise. It seemed apt. In Watson's war on the whalers, there were no conditions for truce.

Aultman spiraled the chopper to 600 feet and we raced over the puzzled, frozen plain. With no references it seemed the pack was just below our boots. We climbed and broke north over open water. We ran straight into the gleam of low sun. And then I saw the ship, a black shadow on the dark blue sea that curved to the horizon. She was pitching into the swell, raising a bow wave.

Conditions had worsened since we left. We had been gone over three hours, the longest flight he'd yet taken, and I suppose we were lucky that the seas hadn't become even more raucous. Had the *Farley* been rolling this hard before, Aultman would never have taken off. He lowered the helicopter to the tossing deck twice, hovered, then balked and jammed the throttle and lifted again for a go-around. He got on the radio. "Hey, Alex, how fast are you going?"

"Two knots."

"Could you increase it to four? And keep her as straight as you can into the waves."

"Sure thing."

The ship smoothed out a little. Aultman brought the chopper down to within four feet of the deck and hovered there like a bee for a full minute, gauging the wind and the motion of his landing surface, which was rolling and yawing and pitching all at once. His right hand on the stick was in motion, adjusting, adjusting. And then he dropped it. Suddenly. Fast but not hard. The deck crew rushed forward and slung the strap hooks and ratcheted the chopper down to keep it from sliding off into the water. They took only seconds. Pretty good for vegans with advanced degrees. Alex pitched the *Farley*'s prop for more thrust and called Trevor for full power and we were under way again, as fast as the *Farley* could go, due west. Away from Australia, away from a return route we could make with the fuel we had left.

At midnight, Watson got a report from his mole on the *Esperanza:* the *Nisshin Maru* and one harpoon vessel were now 650 miles to the west of us. They were staying just outside the 200-mile limit of the Australian Antarctic Territory, and continuing west at fifteen knots. No sign of the spotter vessel or the three other killer ships.

18

The Definition of a Pirate

December 31, 2005
64° 16'S, 125° 22'E

The last day of 2005 was unfolding without incident. Sometime in the night we had passed north of Porpoise Bay and Cape Goodenough and left the Clarie Coast behind. We were heading west. That was all anybody knew. Trevor kept sounding the fuel tanks, and he looked less and less happy. Were we to run out of diesel we would be no better off than a ship with a tangled prop. Should a big gale come up, like the one from which we had just emerged, we would flounder and beat against the floating ice and sink.

Last night at midnight Watson got another communication from M—on the *Esperanza*. The *Nisshin Maru* had the one harpoon boat with it, and the two vessels were continuing west at fifteen knots. They were not whaling. At the time of the report, they were at S 61° 02', E 106° 00'. At eight this morning that was 570 miles west of us. M—said the *Arctic Sunrise* was also flagging badly and was now 190 miles behind its sister ship. It seemed we had no hope of catching up unless the *Nisshin Maru* reversed direction or stopped dead in the water. Nor could Watson be hopeful of contacting a harpoon vessel even if we saw one, as they go something like twenty to twenty-five knots. Was he waiting for the possible arrival of a Japanese warship?

The ship was Watson's primary weapon—that and the blitz of press releases—and we all wondered what he intended to do with it. Hatanaka's talk of piracy was disturbing. On a website posting yesterday, Watson said:

> The Institute of Cetacean Research has made an open accusation of piracy and ecoterrorism against the Greenpeace Foundation and the Sea Shepherd Conservation Society. If Japan adopts the false accusations that acts of piracy have been committed against their ships, they can use the accusations as an excuse under international law to attack and seize the ships they accuse. . . . In effect all the Japanese have to do is decide that Greenpeace and Sea Shepherd ships are in violation of Article 101 to intervene. . . .

They could, by rights, shoot their own high-powered rifles into the bridge and the hull.

I went to the little bookshelf in the chart room and pulled down the *UN Convention on the Law of the Sea.* The articles on piracy had been well thumbed:

Article 101
Definition of piracy
    Piracy consists of any of the following acts:
        (a) any illegal acts of violence or detention, or any act of depredation, committed for private ends by the crew or passengers of a private ship or a private aircraft, and directed:
            (i) on the high seas, against another ship or aircraft, or against persons or property on board such ship or aircraft;

(ii) against a ship, aircraft, persons, or property in a place outside the jurisdiction of any State;

(b) any act of voluntary participation in the operation of a ship or aircraft with knowledge of facts making it a pirate ship or aircraft;

(c) any act of inciting or intentionally facilitating an act described in subparagraph (a) or (b).

Article 103
Definition of a pirate ship or aircraft

A ship or aircraft is considered a pirate ship or aircraft if it is intended by the persons in dominant control to be used for the purpose of committing one of the acts referred to in Article 101. The same applies if the ship or aircraft has been used to commit any such act, so long as it remains under the control of the persons guilty of that act.

Article 105
Seizure of a pirate ship or aircraft

On the high seas, or in any other place outside the jurisdiction of any State, every State may seize a pirate ship or aircraft, or a ship or aircraft taken by piracy and under the control of pirates, and arrest the persons and seize the property on board. The courts of the State which carried out the seizure may decide on the penalties to be imposed, and may also determine the action to be taken with regard to the ships, aircraft, or property, subject to the rights of third parties acting in good faith.

You didn't have to be a lawyer to make the case against Sea Shepherd. Welding a seven-foot blade onto the bow for the express purpose of damaging the hull of another ship, and then ramming—or attempting to ram—that weapon into said hull was clearly not legal and could be construed as an act of violence. As could running out a mooring line in an attempt to foul the prop of another ship in a storm. The *Farley*, it seemed, under the Law of the Sea, could now rightly be said to be a pirate ship, subject to attack and seizure. She didn't even need to fly her Jolly Roger. Subparagraph (b) then made any voluntary participant in the operation of the *Farley* a pirate.

The leaden sky closed in around us again; the wind died; we cruised west. Flurries of snow came and went, as did pairs of storm petrels, the small black birds with the white rumps. I went to the green room in the afternoon, and Allison put on a video made by Mark Votier. He was the one who had been hired in 1996 by the Institute of Cetacean Research to make a documentary about Japanese whaling. Votier was so disgusted by what he filmed, that he released the footage. The video was not long, maybe twenty minutes, all taken from the deck of a harpoon boat, or from the *Nisshin Maru*. It had a jerky, home-movie quality, the handheld effect of a cameraman trying to stay on his feet in heavy seas, trying to stay out of the way of flying flensing knives on long poles.

Here was a heavily dressed hand loading the harpoon gun with the explosive-tipped harpoon. Here was the ship moving fast in the swell. There were the blows of a pod of minke whales ahead, fleeing for all they were worth, blowing every few seconds, clearly panicked. *Fire*. The flight of the harpoon, the arrow-straight line of cable following. Miss. Fleeing whales. Now the camera focused on a whale in the rear of the pack. Good size. The harpooneer focused on her too. *Fire*. Miss again. And another. Fourth shot hits her in the flank. Explosion and fountain of blood. Whale thrashing. Cable winch engaged, thrashing, screaming whale reeled in, gushing blood, turning the sea red. Hauled to the side. Still convulsing, hemorrhaging everywhere. Another spear, probe, on long pole with cable attached thrust into her side. Whale writhing. Big generator on deck blaring. Electrocution current now coursing through the new spear. Whale in bloody agony. Not even close to dead. Finally hauled tail up, suspended so they can hold her breathing hole under. She drowns after fifteen more minutes in a sea of her own blood.

I wanted to vomit.

Transfer to mother ship. Two whales being winched by the tail up the slipway. The banner on the stern, "Greenpeace is a sham." A hard-hatted crew stepping forward with the long flensing knives and slicing the first whale open. Three-foot-long fetus removed. Still alive. Tiny scale model of a whale. The pregnant mother was not as fast as the rest of the pod. Fetus carried to a bench with scale, measured and weighed by the scientist. Guts spilled on deck by the long hooks and knives.

The choppy film ended. Silence, but for the reassuring vibration and throb of the diesel engine and chortle of the sea along the hull.

There is no more barbaric method of slaughter on earth, in any meat industry. This prolonged butchery and torture are reserved for the most intelligent, most social order of beings. The fact that the Japanese thought that filming all this would be good PR seemed to point to an almost insane departure from reality.

I went out to the main deck to digest what I'd seen. I leaned against the rail and listened to the soothing rhythm of the bow wave. When I needed to get my physical bearings I pulled out my GPS, got a fix on three satellites, and knew in a moment my exact latitude and longitude. I knew, within a few meters, where I stood. But where did I stand with whaling and Watson's aggressive methods of protection? Out here, my moral bearings were less easy to find. Many people, and nations, were calling Watson a pirate. But was he? Pirates tended to kill people. Watson, on the other hand, was trying to preserve all life, human and animal alike. He said again and again that he wished no one harm. He would damage property alone. And yet his tactics, especially as executed by his freewheeling and inexperienced crew, put people at grave risk.

I thought about something Watson had said earlier. He quoted Jacques Cousteau as saying that the oceans are dying in our lifetime. I watched the wind spray the whitecaps downwind and thought about the first time I had been awed by what I had seen beneath the surface of the sea. I was twenty-three, just out of college and traveling in the Pacific Rim. In Fiji I signed up to go snorkeling off a barrier reef. On a sunny morning we anchored in twenty feet of glass-clear water. I went over the side and kicked toward a ridge of coral. For a few moments I forgot to breathe. Orange and purple coral fans waved with the passing swell, black urchins bristled out of pockets in green and red coral heads, branched plants and coral swayed like forest leaves. And the fish. I swam into schools of brilliant fish and divided them like a ship pushing through currents of color. Yellow fish, speckles of electric cerulean blue, greens and vivid indigos I had never seen. And through it all the sunlight sprayed and the shadows shifted and slid back with a rhythm that was wholly strange to me, and magical. I marveled that the littler fish did not seem alarmed when a bigger, blunt-headed hunter, half as large as me, drifted among them. Could anything in the entire universe be this unabashedly beautiful?

The other time I had forgotten to breathe was when I was in the Sea of Cortez, crewing on a sailboat. We had anchored off some rocky islets that were occupied by an active and voluble sea lion rookery. You could hear the barks and roars of the bulls for a mile. I put on a shortie wet suit and jumped in wearing mask, snorkel, flippers. People had told me that the sea lions here were fairly used to snorkelers and though they would never let you touch them, they would swim by very close. They did. Adult females swam straight at me and twirled as they passed, keeping their eyes locked on mine as they whizzed by. Big bulls ignored me. Curious youngsters jetted by like fast planes. Then I saw two adolescents playing with a shock cord nearby. Some boat had dropped it. One picked it up off the rocky bottom and the other would scream into a dive and snatch the other end and they'd play tug-of-war. I wanted to play too. I took a big breath and dived to the bottom, and as I did I executed an upside-down triple lutz, somersault, jackknife thing. Very awkward. Their heads turned. They looked incredulous. What the hell was that? they seemed to say. They rushed over, and as they did, they each performed a twisting barrel-roll figure eight. Oh, yeah? I responded, Watch this. I dived again, somersaulting and somersaulting, then twisting and fully extending. They were astounded. They sped up, jetted around me like the Blue Angels—loop the loops, flips, twists, rolls. I doubled my efforts. They seemed to think it was hilarious. Then I felt a hard tugging on my flipper. I twisted around to look and one was pulling at my fin and watching me with bright liquid black eyes. Slowly I arched myself into a half circle and grabbed her two little black back flippers. Ever so slowly we turned in the water in a circular doughnut. Then she released and broke away. But she swung around and came back and I reached out my arms and she swam up to my extended hands and took my left fist in her teeth. She had pulled hard on the flipper, but her needle teeth held my fist gently and her black eyes, a foot away, looked directly into mine. The surge of warmth and glee that went through me—I'd never experienced anything like it. We played and played. I held her torso and she pulled me through the water. We hung upside down a few feet apart and just looked at each other. I forgot my name. I forgot to breathe and came to the surface gasping.

When finally I was blue with cold and the crew was calling me and I swam back toward the boat, she swam around and around me in fast figure eights.

If the oceans are dying in our time and we kill them, which is what we are doing, we shall have committed a crime so heinous we shall not ever be redeemed.

That night we had a New Year's party. Mathieu produced the stomped grape juice that had been fermenting in the aft hold. iPods with attached speakers came out, and the mess echoed with grunge and Straightedge. Emily Hunter danced up and down the narrow aisle while Mandy relinquished herself to Dead Head embrace-the-universe twirls. Stashed beers and wine came out. Casson retreated into the galley and rigged up an elaborate still—fermented pear mash in a rice cooker with a taped lid, sending the boil-off through surgical tubing into a beaker. In the end, they just got juiced on the mash. Gedden, the J. Crew, and I played a hot game of Texas Hold 'em in the corner. We all counted down the life of the old year and sent it packing with a yell. New Years is always a funeral and a birthday at the same time.

I went down to my cabin and crawled under the old flannel sleeping bag which was my issued blanket. The lining of the sleeping bag had a repeating print, mostly red, of a canoe on a river, fir trees, a man fishing with a wicker creel, a man in red cap and a dog, the man shooting a flushing pheasant. Why hadn't I noticed that before? The irony was perfect. Some vegan crew should have sewn in the figure of a hunt saboteur banging pots and pans. I liked the scene. It was nostal-gic and comforting. I lay back in the dark, the fluorescent light from the hall outlining the curtain in the doorway, and watched my glow-in-the-dark Milky Way, and felt so lucky that I had grown up in a time when nature had seemed mostly untroubled. I had spent summers in the Adirondacks and on the eastern seashore when the woods and mountains seemed wild, when you could fish from Jackson's boat-house and catch stripers and bluefish, and gather mussels with your grandmother from the rocks at the end of the beach and eat them for dinner and canoe the Saranac and Ausable lakes, and drink directly from the brooks. It was only three or four decades ago that I did all that, but it seems quaint and anachronistic now. Few were worried then about global warming, the collapse of every fishery by 2048, the ocean's being too acidic for coral, or the death of most of the earth's coral reefs and thus perhaps the death of the million-plus species they shelter. Nobody saw then, or very few, the death of the earth as we know it. Of course, we should have known.

19

Among the Penguins

I was up early on the first day of the year and saw no one as I took a cup of steaming coffee out onto the empty main deck. We had converged with the ice edge again and were cruising a few hundreds yards off it. The sea was lake calm and mirrored a scudding sky. The first beings I greeted in 2006 were an impossibly wide-winged sooty albatross and a little snowy petrel, who circled and circled and crossed each other. The albatross was dark brown and came over the deck like the shadow of a flying dinosaur. The petrel was pure white and small, his wings also held straight and still.

Watson was on the bridge, in his chair and uncharacteristically reflective. We were making a steady nine knots west in the calm sea. Finally he spoke up.

"It's just strange. The Japanese have not been seen whaling since December 24. They were idling at six knots in the storm when Sea Shepherd showed up, then ran—fourteen, fifteen, sixteen knots west and haven't stopped. They are outside the Australian economic zone. I can only conclude that they are being chased out of whaling and are either, one, waiting for a Japanese warship; two, waiting to run us out of fuel; or, three, whaling with the three missing harpoon boats and the spotting vessel. Maybe the spotting vessel is also a processing ship. I don't know."

We were now off the Sabrina Coast. He said we would head for Cape Poinsett. On the other side of it was the old Wilkes Station, established by the United States in 1957, later abandoned. We were running low on fuel. Maybe we would wait there, in the shelter of the bay, to see what the Japanese had planned.

A few hours after lunch the *Farley* slowed and approached a lovely, cleanly sculptured iceberg the size of a house. Its edges were sharp as cut crystal and shadowed blue, and on its upper slope a flock of antarctic petrels nestled against the packed snow. Icicles fringed every overhang, and several outriding, smaller floes cupped a protected lagoon in which two large humpbacks idled and blew. Hammarstedt throttled back the engines and we could hear the whales. They were perhaps fifty yards away. On the other side of the ship was a little raft of ice. Out of nowhere, probably attracted by the sound of the engines, a squad of a dozen Adélie penguins skipped like stones over the water and slithered up onto the bobbing floe. They waddled to the edge and dived off. They flipped, flashing their white bellies, then logrolled fast like birds in a birdbath. They wriggled back up on the ice to do it again. They glanced over at the *Farley* and squawked at each other, and waved their flippers, and then ignored us.

The word went out. Trevor announced that the water was thirty degrees Fahrenheit. Crew began to strip on deck, down to their Skivvies. Geert took it all off, displaying his tribal tattoos. They climbed up on the rail and dived in. The penguins sensed the splashes and looked over as one. What the hell? Happy New Year. I stripped to shorts and dived in, too.

The cold was more like an iron club than enveloping water. Was that all my breath leaving in a rush? Is that pain or just the pressure of imminent death? I flailed for the rope ladder and climbed out like a sodden cat.

Darren, shoulders hunched against the faint breeze, handed somebody his glasses and said with resignation, "Well, it's mandatory, isn't it, sir? It is the New Year." He went overboard. But there was a slight current working against the idling *Farley*, and he shot out his arms and swam but lost ground. Little by little he was sliding aft. He had seconds before he would sink. Allison, without a thought, reached behind her, grabbed a coil of deck rope, and shot it to him and pulled him in. Yup, a warrior. Gedden dressed in a full wet suit and jumped off the crow's nest, about fifty feet off the water. Quiet, slow-eyed, J. Crew Joel went unadorned off the bridge wing. Aultman took the chopper up for a short recon flight and returned with no news.

Happy, invigorated, we left the astonished penguins and the lolling whales and continued toward whatever the captain had in mind. Watson went to the radio room and printed out scroll-edged Penguin Certificates for the dozen who went into the water.

The elevated mood didn't last long. There was a cry from the crow's nest. Luke, the grunge rocker, and Simeon had gone up there to take in the view.

"Dead whale!" they cried. "Dead whale! One o'clock!"

The *Farley* slowed. I went out the heavy hatch aft and climbed up to the monkey deck. A shape floated ahead, marked by a flock of birds as if by buoys. In life, in the water, unless it leaps a full-body breach, one never sees the whole body, head to tail, of a whale. Now, in death, he was perfect, except for the raw gouges where the birds had begun to feed. It was a juvenile sperm whale, very young. He was perhaps twenty feet long. His tail was white, the lovely flukes sunk just beneath the surface, as were his thin lower jaw, agape, and his extended penis. Otherwise he was a light, mottled brown. The birds had already taken the eye. He floated in perfect profile on the night blue water. On his flank were half a dozen giant petrels, wings spread in defiance or ready-flight. Fifty more bobbed in the water beside him, along with a few Cape pigeons and snowy petrels that rode respectfully back. As we closed the gap, the alpha birds dropped off into the water—all but one, which hopped up to the whale's square head, gaped, and glared at the ship that was now abreast and looming. He spread his wings and ratcheted out a strangled cry and then retched onto the whale's head, a gobbet, red, which slid into the water and floated away. This is mine, he seemed to say. Back off.

The young whale showed no signs of trauma or injury. He was surely still of nursing age. What had probably happened involved the death of the mother and the slow, stricken dying of the dependent son. The crew, of course, thought immediately of the Japanese. The missing harpoon vessels had probably recently come west along the coast, shadowing the *Nisshin Maru*, but how far out was anybody's guess. They might have come this way. Sperm whales were not on their permit, but that didn't seem to stop the whalers, judging by spot tests of whale meat at the Tsukiji market in recent years.

The crew lined the rails in silence and paid a silent homage to the young whale. Then Watson pitched the prop, and the *Farley* gathered what speed it could muster and we left the whale and the birds behind. We pointed toward the smudge of sun low over the horizon.

Trevor came onto the bridge. It was the second day of January. The *Farley* had been pushing for the last two hours deeper and deeper into a maze of packed floes and broken field ice, trying to get south to Wilkes Station. The fog had rolled in again. Dense, wet fog that beaded every surface, dripped off the safety lines and rails. You couldn't see more than a few hundred yards. Aultman stood with binoculars peering ahead and gave Lincoln directions at the helm. Darren called out growlers. The black leads of open water snaked and zagged into the white world ahead and the *Farley* followed. Four fin whales surfaced just off starboard and kept us company for a few minutes, swimming under the boat. Their flukes were as wide as two vegans. The open leads forked and forked again and then closed down. From five knots to two, pushing through the broken layer. We were surrounded, closed in. Congealed ice ahead, and growlers beginning to fill the open water of our wake. Watson, watching from his chair, turned to his nephew. He'd asked him to sound the tanks one more time.

"Well?"

"Eighteen to twenty-one days of fuel left. That's down to fumes."

"It's 4,000 miles to South Africa, which is about nineteen days—Lincoln, turn back west. It's getting worse."

Lincoln did. The *Farley* swung to starboard in an agony of bumping and grinding.

"Second thought, shut her down. We might as well wait until we can get a helicopter up and see where we're going."

He turned back to Trevor. "OK, thanks. We'll be OK, unless we get a storm or something. Then we'll be hitchhiking in the middle of the Indian Ocean."

The engine shut down. The sudden silence hammered the point home. No sound but the occasional brush of ice against the hull, the trickle of the bilge pump, a thin stream of water into the congealed sea, the ship's life. We drifted. No wind. Our breaths on deck trailed northwest. We were the loneliest ship on earth. The latest communiqué put Greenpeace and the two Japanese ships 650 miles to the west. We were as far as we could get from any inhabited continent, almost equidistant between Australia and South Africa. There was ice closing in on every side.

We drifted with the floes. At lunch Wessel, who had just called home and talked to his mum, said that Sea Shepherd, the whalers, and the Japanese navy were all over the news in South Africa. So it was working. Watson said he had heard again from the *Esperanza* and the fleet was in the east nineties now, not whaling.

We lowered the ship's sea kayak into the water. There were enough cracks and small leads in the pack to maneuver the narrow boat, though there was a danger of getting closed in. Emily borrowed my dry suit and went for a short paddle in the mist. I saw her just sitting for a long time, a quarter-mile from the ship, looking off into the fog. Six months was not very much time since losing a father. I wondered when she would spread his ashes.

When she returned, she was radiant. "Thank you," she said. "That was one of the neatest things I've ever done."

That evening the fog lifted. It was late, after 2000, when Watson sent Aultman up to search out a route. He came back two hours later and said there was ice to the south, as we knew, and a big peninsula of ice to the west. He said that to get to Wilkes, or west at all, the route was northwest, up and around it, then back south.

Watson waited. He pondered his next move and let the *Farley* drift. Every hour he ran the engines was an hour closer to empty fuel tanks. He didn't want to waste a minute running the engine without a plan.

Trevor, without much to do for once, fished out a hunk of growler, broke it up, and poured himself a rum and Coke on the rocks. He stood at the stern rail in his coveralls watching the world go by—ever so slowly. The port side of the ship nudged up against a small flat-topped iceberg, about the size of a boxing ring. Its top was about five feet below the level of the deck Trevor stood on.

"Do me a favor," he said, turning to Dennis. "Hold this for a second. Drink it, you die."

He handed the deck hand his drink, then vaulted over the side. He stood on the berg and waved as the ship edged forward. Lincoln thought that was cool. He jumped over, too. The ship slipped farther ahead. Trevor took a running start, jumped, grabbed a bottom rail and hauled himself aboard. Lincoln judged the distance, but it was too late. A four-foot gap had opened between the two platforms. Lincoln stood alone on his little berg looking like the Little Prince as the ship, on the surface current not affecting the deep-bellied iceberg, began to leave him behind. Trevor and Wessel, acting fast, went to the big spool of longline on the stern. They ran off 100 feet and tossed Lincoln the coil. Just in time, he tied a tight bowline around his waist, leaped off the ice, and dangled two feet above the water, and five of us hauled him aboard.

That night I stayed up late watching a DVD of the television show *Nip/Tuck* in the green room. It seemed to be about morally rudderless rich playboy doctors in Miami. There was a lot of sex and fancy cars. The hardest-core vegans on the ship were the ones who loved it the most. During an intermission, Mathieu and Kristian dug out some rusted old cans of meat ravioli they'd found in the bottom of a bin in the aft pantry, and they opened them and heated them up with gusto, to the profound disgust of Mandy and Inde and Laura. I tried a fork-ful, but it tasted like chunks of mildewed leather.

We started the engines and began moving just before noon on January 3. Not in a hurry. We were at 64 degrees south, 111 degrees east. Watson had just gotten word from the second mate on the *Esperanza*, who had not previously been in contact. She confirmed in an e-mail that the entire fleet had refueled at longitude eighty-five degrees east. That was far to the west of us, over 700 miles; and four latitudes, or 240 miles, to the north. It was also just inside the treaty zone, which was explicitly illegal. She also said they had lost contact with the *Arctic Sunrise*, which seemed strange.

Alex took the helm, muttering about how the Japanese refuel wherever the hell they want, twenty-six miles inside the treaty zone, they just don't give a fuck. We're Japanese, we can do whatever the fuck we want. The quiet first mate was seeing red.

"How do I get out of here?" he said, scanning the ice-choked sea and looking back east toward the only open water. He pushed the *Farley* around. "You know, the probability is that the Japanese fleet is much farther north than we thought. If we go to the abandoned station it's farther south. So if they come back and we're stuck south and get iced in, then what?"

He gave the helm to Gedden and went to the charts with Watson. He said, "If they, the Japanese, hang around that area, then there's no point in waiting here."

Watson pointed to two features on the chart due west of us on the sixty-fourth parallel, about 270 miles, called the Ice Domes. They were two little prickly footballs on the map, ten miles apart. They looked like a pair of porcupines, above the parenthetical notation "Rep 1947." Reported sixty years ago. No mention in *Sailing Directions*. Alex nodded.

"Yeah, we'll head in that direction, and hopefully by then we'll know more from our contacts. If they return east, we can hide there and wait for them, or—whatever."

Gedden zigzagged back the way we'd come, then turned north, hunting open water. At times the engines slowed and the *Farley* broke her own trail, shoving aside floes, cracking open a path through the floating puzzle. It would be so easy to get iced in. She broke through into a black sea scattered with bergs and turned northwest.

We were now roughly eighteen days steady sailing from Perth, if we turned around and went back to Australia. Cape Town was a bit farther. That was the very limit of our fuel, should we not encounter any big storms. Heading farther west along the coast was stepping into the void. *We do what we do because it's the right thing to do, don't worry about the consequences or even the results.* Watson was certainly walking the talk. Alex, Watson, and the other officers were very quiet this morning.

The rest of the crew seemed intent on keeping up morale and passing time. In the afternoon we came on a magnificent island of ice that had been weathered into shapes of swooping hills. A large humpback lolled in the cove, blowing and throwing a long fin in the air. A bunch of hands put on wet suits with Mustang suits over them and headed off in a Zodiac for the edge of the island. At the last moment diffident Darren jumped into the Zodiac. Watson, looking down from the bridge, said, "What the hell is Darren doing in there? He's not wearing a survival suit. Jesus, call them back." Too late. The inflatable buzzed to a low shelf along the bank of ice and, maneuvering with difficulty against the swell, pulled up close enough for Trevor to leap off. He had on crampons and he used a claw hammer and a hatchet for ice tools. I noticed through binoculars that he had a snowboard slung across his back. He hit the steep bank and stuck to it, spread-eagled, like Spiderman. He began to climb slowly. When the slope shallowed he stood and walked carefully. Fifty feet up he slung his tools, threw down the board and rode it on his knees straight down the berg and off the ice shelf.

Splash!

Watson blew the ship's horn. The whale lifted his flukes and dived.

When Darren clambered back out on the main deck, welder Marc said, "Welcome back aboard. No Mustang suit—you're sick."

Darren wheeled around and snapped, "At least I'm not sexist and homophobic!"

Touchy. It startled me—the remark came out of nowhere. The uncertainty was wearing on the Shepherds.

■ ■ ■

Watson was heading west. I had told my fiancé, Kim, that I would be home the first week of January. Which was now. I missed her badly. I longed for home. I missed feeling useful. We were at the southern reach of the southern Indian Ocean halfway between two continents on a ship that could go no faster than a man could run. Wherever we were going, we weren't going to be anywhere near an airport for weeks—if we got there at all.

Maybe the most dangerous enemy out here was not a monstrous factory ship, but the monotony of vast expanses of water. The frustrating lack of action. As the gap between the fleet and the *Farley* increased, the sense of our effectiveness was slipping away. Yet the whaling fleet had not killed a whale for nine days. It was not hunting but running to the other side of its research area. However fruitless the chase at the moment, Sea Shepherd seemed to be having an effect.

20

A Good Day to Die

Sometime in the night we rounded the corner and turned again due west. When I woke there was a muscular new swell out of the east and the waves were whitecapping. January 4. The barometer was dropping. No telling if this was just a freshening breeze or the beginning of a new storm. Nothing to do but go with the seas and motor now that the decision had been made.

I wondered if Watson was getting frustrated. He didn't show it. The day passed quietly and the crew kept busy. Chris Price tried to convince everybody that the assassination of J.F.K. was the fascists consolidating their power. Geert told him to turn around and traced a whale-tail design on his buzzed stubble; Kalifi shaved around it, so the trucker had flukes of gray fuzz on the back of his skull. Jeff worked on a plywood heart edged with nails and scarred with incisions, which he sewed up with twine. "I'm giving it to my wife. What she did to my heart. For Valentine's," he said.

At 2220 Watson said, "I just got a new position on the Japanese. They're at sixty-two degrees south and seventy-three degrees east, heading south. They've gone almost to the opposite side of their area—700 miles west of us. Haven't killed any whales yet. I think they've probably done this on purpose to get as far away from us as possible. And Greenpeace doesn't seem to know where the *Arctic Sunrise* is. It's the second time they've asked me."

Alex went to the computer nav station and typed in the coordinates. He then quickly plotted a course to the Japanese fleet and back to Perth.

"Three thousand two hundred miles."

"Same as going to Cape Town, isn't it?" Watson said.

"Pretty much. At seven and a half knots it'll take us eighteen days from here to the whalers and back to Perth. One extra day to Cape Town, actually."

"Keep going," Watson said. "We need one more hit at them." He went below.

Dawn on January 5 was dark and windy, with whitecapping waves. It was almost like a true dawn, the kind that comes after a black night. But there had been no black night; there were only the cloudbanks thickening and growing heavy and dark with snow. The sky massed into a seamless monolith that dimmed the air and lowered almost to the water and turned it black. If it moved, it moved against the sea in one implacable march, without division, trailing paler wisps like banners. We were far enough from the ice edge now that the only ice we saw were lone drifting islands that thundered and boomed with the swells. The icebergs met the seas with cliffs of alabaster and the waves smashed against them and hurtled spray over their tops. On the black water, the snowy bow wake of the *Farley* and the ripped combers of the whitecaps were a relief to the eye. We had been trailed all night by two black-browed albatross that seemed still to consider us worthy company. I went to the chart and figured they had been following us for 110 miles.

We had passed the Ice Domes, if they still existed. Watson had only a bare minimum of fuel and he was going to try to use it to attack and figure out how to get home later. The latest idea was to make a beeline for the Kerguelen Islands, a lonely group on the forty-ninth parallel and not too far off the route to Cape Town. Watson said there was a

French base there—maybe we could buy fuel from them. Or call a barge to meet us 1,000 miles out of Cape Town. Trevor said he thought we could make it. He said he had a few tricks. "I can crawl into the tanks and extend the suction lines, get three days more fuel. It's dirty, I'll have to clean it a few times. I'll stand by whatever decision the captain makes."

Trevor stopped in the chart room and tapped a pencil on the nav computer screen. The *Farley* was a blue wedge-shaped blip in the middle of nowhere. "If we run out of fuel off of Africa it's no problem. There are a shitload of boats off Africa. Just call in a barge to bring some out. If we have trailing seas the whole way, we're singing. Then again, the weather through here"—he moved his pencil across the forties and fifties—"is pretty much always shit, as we saw." He squeezed my shoulder. "We'll get somewhere, that's guaranteed." He went around the corner and disappeared down the steps.

The wind and the waves pushed us all day long. We moved west in a choppy following sea and the sun must have made its low half-circuit, but we never saw it. After the false dawn there was a false dusk, and we moved through it hour after hour.

Watson distracted himself on the bridge by reading aloud from another one of his books in progress, *All the Dope on the Popes*. "What I did was give each pope a grade-point average and a collective GPA on every 250 years to show the decline of the papacy." When he was done with that, he read from one of his favorite books, *The Darwin Awards*, true accounts of the most idiotic ways people die.

"Tell them the one about the ice," Allison said.

"Oh, this guy threw some dynamite out to make a hole in the ice for ice fishing and his dog fetches it, runs under the truck. All died."

There was a reverent silence, for the dog.

"Poor critter," Mandy said. I thought Inde was going to cry.

Again, at 2200, Watson received a report from his mole on the *Esperanza*. Now he didn't look so pleased.

"We've got a new position. They're killing today. They started this morning."

He said they were 510 miles due west. The gap was shrinking.

"At seventy-three degrees east. They've gone from one extreme edge of the boundary to the other: 3,000 miles. They're running from something. I don't think it's Greenpeace. I have to put on the website that we're going back to Perth. Surprise them. We don't have to hang around, we just have to get them. We don't have enough fuel to fart around."

"Do we have enough fuel to not fart around?" Lincoln asked. Good question.

"Two and a half more days and we'll get them. Maybe we can talk Greenpeace into giving us some fuel. Probably not. I think we'll probably go to Cape Town. They can't go much farther. They're at the edge of their boundary."

The weather held and the *Farley* made good time through the night of January 5. All night it plowed under the same lowering sky. But the ship woke up. Whatever desultory mood had gripped the crew was gone. The word had passed that we were two days away and that the captain meant to engage at all costs. Out came the grinders, the steel cable, the buoys, and line weights. This was their last shot.

On January 6, we woke to discover that the U.S. Office of Naval Intelligence (ONI) had placed the *Farley Mowat* on its piracy watch. The news was on the international wire services. Online, I found the ONI report listing Sea Shepherd in a communiqué of the Civil Maritime Analysis Department titled "Worldwide Threat to Shipping Mariner Warning Information." The ONI would continue to monitor the situation. Watson's mole on the *Esperanza* reported thirteen minke whales and one endangered fin whale killed yesterday. The Japanese had killed an endangered whale in a whale sanctuary in contravention of at least half a dozen international laws, and the United States was responding by putting the screws to Sea Shepherd. In New Zealand, the whaling controversy was the top story on television news, and its minister of conservation, Chris Carter, said, "The program the Japanese are undertaking in the Southern Ocean is not about science, it's about hunting and killing whales to supply meat markets."

He added that New Zealand would be upping the pressure on Japan to stop whaling, and that the New Zealand air force would be sending Orion surveillance aircraft to monitor Japan's activities. Everybody was being monitored by somebody.

I went out onto the main deck to clear my head. The skies had lifted and the seas had calmed to a gentle roll out of the southwest. Then I saw it: a plume of mist just off the starboard bow, and another, smaller. A white-mottled long fin gestured out of the water, and the two glossy dark backs, mother and calf, dived under the boat, the mother's articulated, graceful fluke disappearing last. Farther off starboard were three more blows, the hot mist trailing gently downwind. Behind them were more and more. Spouts of steam rising and drifting. I stared, almost stricken. All the way to the horizon, where two flat-top icebergs marked the edge of the world, were humpback whales swimming slowly east. Pairs and small groups rolled around each other, showing fins, flukes, eyes, and then moved on. They swam past the boat on both sides. Hundreds of whales. Could they be swimming away from their hunters in the quadrant to the west? They were not on the Japanese quotas until next season, but no sizable whale was safe down here. They were not concerned with us at all. I tried to imagine the migration generations earlier, when they weren't a fragmented, isolated population, when they numbered fifty times what they are today.

Alex called a crew meeting. He said we were a day and a half from the whalers and that we would engage them. He said the new Zodiac crews would be posted on the board. "We are going for maximum damage. If you try to steal their banners I will personally pop your pontoons," he said. He would also post battle stations for every crew member. No one would be forced to stay below. He told everybody to grab a Mustang suit, label it, and stash it in the green room for fast access.

Within minutes the *Farley* was a hive of activity again. The sweet burnt smell of grinding metal wafted out of the bosun's locker where more prop foulers were being hastily manufactured. Others carried them out and coiled the cables and stashed them in the inflatables.

Crews cleared and cleaned the decks for fast access to the lifeboats, the water cannons. Marc installed his shiny new lifeboat cradle on the aft deck port side, moving it from its awkward spot behind the rail and centering the pod now between two stanchions where it could be quickly shoved over the side. "You don't wanna be fucking around." He turned a new bolt through a bottom flange and whistled. "Drop it over between rail posts, that's it. Every second could save your life."

The companionway to the mess was crowded with crew carrying exposure suits into the green room. The vomit-like reek of butyric acid from the old suits polluted the room. Trevor dogged down the forepeak, sealing all hatches and bulkheads forward of the fish hold. He said the heavy bow could be destroyed in a collision and the ship would still float. People were now taking very seriously the possibility of a sinking, of being dumped into the antarctic drink.

On my way out through the chart room I noticed—oddly, for the first time—the paper from a fortune cookie in the tiniest wood frame tacked up on the bulkhead. It said, "☺ life to you is a dashing and bold adventure ☺ "

That evening Aultman lifted off with Watson and Emily as passengers, the first time he'd ventured three in the chopper. They were going back to a large flat-top we'd seen on the southern horizon. Emily carried the box of her father's ashes. She told me later that they landed right on top of the tall berg, got out onto the packed snow, and performed a ceremony. It was an appropriate last good-bye. Hunter had been Watson's first comrade in arms protecting whales; they had been together in the Zodiac the first time anyone had ever run an inflatable between a harpoon and a whale; and they had both looked into the eye of the big male gray whale that had pulled away from crushing them as he died. Emily read something and then commended her father to the ice.

On the way back to the ship they saw a long fin whale swimming with her calf. The pair churned the water to a lighter blue. They surfaced and spouted, one after another, the larger cloud of mist followed by the smaller.

Before dinner the sun broke through the cloud deck and silvered a great swath of sea to the south. And the seas were perfectly calm. It seemed an omen. Only the second time we'd seen the sun in a month. In the packed mess, just as the crew was digging into vegan chocolate cake, Chris Price announced that Sonny Bono had been killed by the CIA. Forks and spoons stopped.

"That's ridiculous," Watson said, standing against the video shelf, chowing down on his second dessert. "Sonny Bono hit a tree. He died skiing."

"That's what they tell you," Price said.

"That's absurd. He was one of them—a conservative Republican."

"That's what they tell you."

The crew started lobbing crumpled napkins at the big trucker. It looked like a hailstorm.

January 8 was a good day to die. For the crew of the vegan pirate ship the *Farley Mowat* that was the consensus. Before midnight the crew of the *Esperanza* had sent another e-mail saying that the fleet had moved south to about 65° 20′ S, 72° 40′ E, so we turned southwest and kept her steady at nine knots and I think everybody held their collective breath. The morning brightened over a sea of silk and glass. The port and starboard Zodiacs looked clean and ready, the tarps off, the prop foulers coiled and stowed. The Jolly Roger was raised and flapped lazily over the bow.

Aultman had gotten off at 0600 and returned two hours later. He was amped. He jumped out onto the heli deck, pried off his helmet, and went straight to Watson and the chart table.

"Here was when we first saw them, twenty miles off our nose. Saw a total of four ships. The factory ship stands out like a sore thumb. They're sitting dead in the water. No wake. I didn't see any activity at all. No lights. There's no wind." He glanced up at the GPS. "The course you're on should put you right on them. Paul, you would not believe the krill blooms. Like a pink checkerboard."

At ten-fifteen a.m. the main radar was lit up with ships. At ten-fifty-nine the factory ship was visible, long and dark, lying across the mirrorlike water. It wasn't moving. The *Esperanza* was there to the south of it, hanging some distance away. We came out of the east at 9.5 knots, heading straight for the target. At eleven forty-seven, with the gap narrowed to under four miles, an irate Scandinavian-accented voice exploded out of channel 16: "You idiot! Read the rules! Read the rules! Get out of the fucking way!" It was the usually courtly Arne Sorensen, captain of Greenpeace's *Arctic Sunrise*. Then we saw a curious thing. From behind the huge factory ship appeared another ship, also black-hulled and just as big. It moved away to the south, to our left. Two harpoon vessels, on their way in when the *Farley* showed up, slipped out of radar range.

What we found out later was that the big tanker-freighter was the cargo ship *Oriental Bluebird*. It had been tied up to the *Nisshin Maru*, and they were in the middle of transferring whale meat from the factory ship for transport back to Japan. Since the Japanese had doubled their quota from the year before, they did not have enough room even on the vast *Nisshin* to store the tons of meat they were harvesting. It was a startling testament to the scale of the hunt.

Greenpeace was bearing witness to the transfer, and protesting by maneuvering the *Arctic Sunrise* near the *Bluebird* and sending in Zodiacs to paint its side with long-handled brushes—"Whale Meat from Whale Sanctuary" in big white letters. The Japanese ignored them and continued loading. Until Sea Shepherd came into visual range. Then they panicked. They dropped the lines, and in the ensuing rush to disengage and run, they rammed Greenpeace's *Arctic Sunrise* twice. An e-mail to Watson from the cooperating Greenpeace crew member on the *Esperanza* said, "Very frustrating indeed. All of this right under our noses, because they know we will not ram or endanger them. . . . At least when you show up, they run like cowards!"

They did. They ran. From the bridge Watson watched the *Nisshin* gather speed and charge north; the *Oriental Bluebird* fled east. From a window on the bridge he yelled, "OK! Get moving!" He slowed the *Farley* long enough to lower the port and starboard Zodiacs into the sea, then throttled ahead as soon as the eighteen-foot rubber boats were unhooked and sped away. Two minutes later the helicopter lifted off the heli deck, tilted hard forward, and accelerated toward the fray. Up ahead we could see Greenpeace's two orange Zodiacs in the water and its orange Hughes 500 chopper circling overhead. It was a melee. At 9.5 knots all the *Farley* could do now was follow and watch as the fastest Zodiac ate up the distance. The *Nisshin* was clearly expecting a fight—its water cannons were blasting steadily over the stern and sides.

"Our Zodiac is going thirty knots" Alex said from the main radar screen.

The second, slower Zodiac came behind it. We watched as the first caught the *Nisshin* and skirted the veil of blasting water. The second reached the side of the ship. Its crew threw two prop foulers against the hull, and then the temperamental old outboard began to balk.

Everybody knew that for a ship the size of the *Nisshin*, there is no way to replace a fouled prop with a spare; the parts are too massive. The Japanese knew that if one of the prop foulers caught and got tangled the crew would have to abandon ship to the killer or spotter boats, and the fleet would have to return home. Wessel was going to do whatever he could to make that happen. Through binoculars I watched as he ran the Zodiac up under the bow of the ship. Even on the smooth water there was a low swell, and the outboard hit those waves and skipped airborne. The prow towered over the little boat, pushing up a bow wave that the Zodiac rode. The Japanese tried to reach them with flensing knives on long poles, but couldn't—they were under the overhang of the bow and too far down. At times Wessel was no more than a man's length from the *Nisshin*'s nose. Joel and DMZ Steve deployed the prop foulers. When they'd used them all, they tied a piece of scrap steel to a buoy with a long cable and threw that over. Then Wessel slid the Zodiac away in a swooping arc and circled to the stern where they picked up the buoy and sped forward to deploy it again. A false move and they'd be flattened under the *Nisshin Maru* like so much roadkill.

Two hours later, lagging farther and farther behind, Watson called the Zodiacs back to the *Farley*. None of the prop foulers had engaged. By three-thirty p.m. the *Nisshin* was sixteen miles away and fleeing at fourteen knots. The fleet had vanished in all directions. The sun began its slide to the horizon, reefs of clouds lowered in from the north and east and the seas doubled in height every hour.

That night, steep swells churned in from the east. They were the outriders of another storm. Whole wheat chapatis and poker chips slid off the tables. Snow blew horizontally across the decks. Sleepers were thrown from their bunks. Hammarstedt, commanding his watch at 0300, went below and woke up Watson, requesting a course change to the east, directly into the seas. Watson approved it, and for a few hours the violent roll shifted to a steady hard pitch. It didn't last long. When Alex took his watch at 0400 he kept going to the chart, getting more and more frustrated that the *Farley* was getting farther from her prey, which had gone north. At 0600 he ordered a course change back north.

Watson came on watch at eight a.m. The snow had stopped. The *Farley* moved gingerly into a sea of scattered icebergs and fog. At nine-oh-five the captain was lecturing the bridge on the fall of Jerusalem during the Crusades when Darren reported a new blip on the screen. Watson swiveled in his chair.

"Range?"

"Sixteen miles, sir."

"Speed?"

"Ten knots."

The fog was so dense it wasn't until we were 1.6 miles away that we could see the black shape extending from behind the island of ice. It looked as if it was trying to hide. I took out binoculars and made out the white letters that I knew said "Whale Meat from Whale Sanctuary" along the starboard side. The *Oriental Bluebird*.

It never moved. Cornelissen was woken up and took the helm. He turned to Kalifi in her sequined sandals. "Tell everyone we're gonna hit. Seven minutes." The crew was ordered to the higher decks and armed with smoke bombs and bottles of butyric acid, the mega stink bombs. Sea Shepherd actions always seemed like a strange mix of a Jack Aubrey attack and *Animal House*.

A few hundred yards from the freighter, Watson took over the helm. He aimed for midships and charged full speed. We were bow to bow, starboard to starboard. Just before contact he threw the wheel over to port so the can opener's seven feet of steel blade on the starboard bow raked the freighter's side. The *Farley* lurched with the impact. There was an agonizing claw-scrape of steel and then another shove as the stern swung in and hit. The can opener crumpled, leaving a long scratch in the *Bluebird*'s thick hull like a keyed car. Watson picked up the mike: "*Oriental Bluebird*, or should I say the SS *Whale Meat*, please remove yourself from these waters. You're in violation of international conservation regulations. You are in a whale sanctuary, and you are assisting an illegal activity. Remove yourself from these waters immediately."

At the same time he swung to port in a tight arc and came back across its bow. The prop-fouling squad was ready to run out a mooring line on the stern and lay a tangler across the freighter's path. By now the *Bluebird* was fleeing. Watson drove the *Farley* within twenty yards of their high hammering bow. It was like running a red light in front of a moving semi. Had the *Bluebird* kept up its speed, they would have T-boned and sunk us. It was as if that's what Watson was tempting them to do. It seemed he wanted them to do it. But the *Bluebird's* skipper, in all prudence, jammed his engines into reverse and groaned past our stern. The deck hands released the fouling line, but it did not get sucked up. By the time Watson could get around again they were running due north at sixteen knots. They ran all the way back to Japan, and never met up again with the *Nisshin Maru* to finish offloading the whale meat.

The international reaction was immediate. The president of the Japan Whaling Association, Keiichi Nakajima, accused Sea Shepherd of being "circus performers" and "dangerous vegans." The newspaper *The Age* in Melbourne reported that Japan was considering scrambling police aircraft to the antarctic to defend its whaling fleet and might ask Australia for protection. The Australian environment minister, Ian Campbell, called Watson a "rogue pirate" and a "lunatic." The Maritime Union of New Zealand announced that it would not service any Japanese ships having anything to do with whaling. Watson dispatched a press release saying that he would stop his attacks if the governments of New Zealand and Australia would initiate legal action to stop the whaling.

But the truth was that Watson had no more attacks to launch. The old *Farley* was nearly out of fuel. Trevor came onto the bridge in his blue coveralls, hearing protectors propped on his head, and said, "We might make it to Cape Town on fumes if we don't encounter too much bad weather." He ordered the heat turned off to save energy. The cabins went cold.

On January 10, The *Farley Mowat* turned away from the whales and limped northward into another gale. There was nothing more she could do. The Japanese could hunt again without intervention. I spent days by the stern chains watching the acrobatics of the petrels and albatross. The rougher the weather got, the more fun they seemed to have. I thought they were like Watson. One image from the trip kept coming back to me. It was when Aultman took me up in the chopper and we ran along the desolate false coast of the ice edge. Coming back to the ship, we climbed to 4,000 feet and ran across open water into a sheen of low sun. Then I saw the ship, a jaunty, compact, black shadow on the taut blue sea that curved to the horizon. It looked completely self-sufficient and alone.

Epilogue

The whaling fleet fled to the west and continued to hunt off the Mac. Robertson Coast. It returned to Japan on April 14, with ten fin whales killed, and 853 minkes—eighty-two minke whales shy of its self-allotted quota.

Three weeks before, on March 24, under international pressure, Nissui and the four other fishing companies that owned Kyodo Senpaku divested themselves of their shares and got out of whaling, citing declining profits. They gave their assets, including the whaling fleet, to public interest corporations, including the Institute of Cetacean Research.

The government and the Institute of Cetacean Research vowed to continue the antarctic hunt, for minke whales and fins in December 2006.

On June 16–20, the International Whaling Commission held its annual meeting in Saint Kitts. Led by the United States, Britain, Australia, and New Zealand, the strong contingent opposed to commercial whaling accused the Japanese of buying votes from small, poor countries. It was observed that the year before, Japan had given $5.9 million to Saint Kitts–Nevis, $11.5 million to Nicaragua, and $5.5 million to the Pacific island nation of Palau. All voted with Japan on all votes that came up. Watson claimed that Togo paid its $10,000 membership fees with a sack full of yen.

With its allies, Japan launched a frontal attack against the moratorium on commercial whaling. Overturning the moratorium requires a three-quarters majority, but Japan, with a simple majority, succeeded in passing a nonbinding resolution that declared the moratorium "no longer necessary" and attributed the declining of the world's fish stocks to whales. The vote passed, thirty-three to thirty-two.

In the November 2006 issue of *Science*, a report by an international team of scientists studying a vast amount of data gathered between 1950 and 2003 declared that if current trends of fishing and pollution continue, every fishery in the world's oceans will collapse by 2048. No more fish sticks. No more snorkeling along reefs with schools of fish. No more fish cat food. No more fish. The oceans as an ecosystem would completely collapse.

A report published by a coalition of federal and university scientists a few months earlier warned that the oceans are now more acidic than they have been for the last 650,000 years, owing to increased greenhouse gases and the sequestration in the sea of more carbon from the atmosphere. The report said that by 2100 the oceans could be more acidic than they have been in millions of years, leading to the death of creatures that secrete skeletal structures like coral, shellfish, and calciferous phytoplankton.

On October 17, Iceland announced that it would resume commercial whaling, in defiance of the IWC. The quotas for the first year would be nine fin whales and thirty minkes. On October 21 the Icelanders killed their first fin whale.

Also, in October 2006, Canada pulled its registry from the *Farley Mowat*. On December 19, now under a Belizean flag, the *Farley* again sailed for Antarctica to hunt down and attack the Japanese whaling fleet. This time, she was fitted with a can opener made from a bulldozer-strength blade. Watson also dispatched from Scotland the newly purchased *Robert Hunter*, a fast, 195 foot former Scottish Fisheries enforcement vessel. The ship could cruise at sixteen knots, one knot faster than the *Nisshin Maru*.

The *Farley* was only hours out of Hobart, when Belize, under pressure from Japan, stripped her registry, making her officially a stateless ship. Any naval vessel of any nation could now attack her at will, seize the ship and detain her crew.

On February 9, after nearly two months of fruitless searching in the Ross Sea, Sea Shepherd found the *Nisshin Maru*. Bottles of butyric acid were tossed onto her decks from the *Robert Hunter*; the *Farley Mowat*, about three miles behind, sent out two attack Zodiacs. The seas were running over seven feet and the Zodiacs were getting pounded and damaged as they sped to catch up to the fleeing factory ship. At a critical moment, one Zodiac began taking on water, a thick fog set in, and the little boat became lost. Watson was forced to abort his attacks and send out an international distress signal reporting missing crew. The *Nisshin Maru* responded and joined the search. Eight hours later the *Farley* found the Zodiac with its two crew members unharmed.

Three days later, the *Robert Hunter* ran down the spotter vessel *Kaiko Maru* in waters beset with ice. Several collisions followed, and Sea Shepherd Zodiacs managed to foul the prop of the killer ship. The *Kaiko Maru* limped back to Japan at five knots with a damaged propeller.

Shortly thereafter, the Sea Shepherd ships, running low on fuel, turned back toward Australia. On February 15, when the *Farley* and the *Robert Hunter* were 1100 miles away from the Japanese fleet, the *Nisshin Maru* suffered an explosion and raging fire below decks. One sailor died. For ten days the disabled ship, laden with 343,000 gallons of fuel oil, floated without power, dangerously close to the pristine coast of Antarctica and the largest Adélie penguin rookery in the world. Japan refused repeated appeals from New Zealand and other governments to avert a potential environmental disaster and have the ship towed out of the area. On February 28, the Japanese fleet aborted its whaling season, and headed home with 508 whales.

On February 13, Japan hosted a three day conference to push for the lifting of the moratorium on commercial whaling. All seventy-two member nations of the IWC were invited. Thirty-four nations boycotted the meeting.

In the 2007–2008 whaling season in Antarctica, the Japanese will begin taking fifty endangered humpback whales, along with fifty endangered fin whales and 935 minke whales.

Afterword

Taiji, Japan

October 27, 2007

Hours before daylight, Charles Hambleton and Louis Psyhoyos lowered themselves hand to hand, tree to tree, off a rain-slick cliff. They dropped to a ledge hemmed by brush, high above a narrow cove. Swiftly, they unshucked two tripods and freed the cameras from their waterproof cases. They worked without light, by feel. I handed down a duffel full of camouflage and followed. I could hear Charles click the FLIR cam onto its mount. Forward Looking Infrared is a military grade thermal video camera, and Charles would use it to sweep the tiny beach and the rocks along the water for patrols. Security here was very tight. Razor wire ribboned across the access trails, and when the killing began the entire inlet would be tarped off. Still working fast, the men tore leaves off the brush behind them and wire-tied them to the legs of the tripods.

It began to rain again, a cold drizzle that blew in from the North Pacific. It dripped from the trees and pattered on the hoods of our slickers.

"The rock-cam is weighted down, right?" Louis whispered.

"Yeah, it should be cool."

A half hour earlier we had placed two other cameras at water's edge. One was in a case fashioned by George Lucas's lab, Industrial Light and Magic, to look exactly like the rocks in the cove. Another was a hot-rodded bird's nest, which Charles hung from a tree. He was the former lead guitarist of the Samples, and I thought he was pretty agile for a rock star. But he was also a master sailor so it shouldn't have surprised me. Louis had been a top photographer for *National Geographic* for two decades and had recently given that up to make this movie. Their company, named the Oceanic Preservation Society (OPS), was based in Boulder, Colorado. The two had read *The Whale Warriors* and invited me to come to Japan. I had gotten an assignment to cover the story from a magazine, gone to the law-enforcement store in Denver and bought an all-black SWAT outfit (Louis had said they would supply the face paint), and booked a ticket to Osaka.

We huddled together in the wind and rain. I told Charles "Feel Us Shaking" was one of my all-time favorites; it turned out he wrote the guitar part. We shivered and ate Louis's chocolate bar. I could hear the swell breaking on the rocks below, and when I stood to shake off the chill I could see a navigational light far off, blinking on some headland. The faint, intermittent spark kept time against the vast blackness of the sea and gave the night a sense of profound and lonely silence.

I was having fun playing commando journalist, but I was also filled with dread. Below us, just across a two-fingered inlet, was the Killing Cove, where 2,300 dolphins and small whales are butchered every year. It's the place Allison and Alex had infiltrated in 2002, managing to cut the nets and free some fifteen dolphins before the two were assaulted by fishermen and arrested. The killing here is part of a cetacean slaughter that is unregulated by the International Whaling Commission, which has no jurisdiction over the smallest whales. Unlike the whaling done in the open ocean, the killing in Taiji is unregulated; the Japanese don't even have to pretend that it's for scientific research. The government issues permits to fishermen, who then kill more than 22,000 dolphins, porpoises, pilot whales, and false killer whales every year along Japan's coasts. The meat is sold to school lunch programs and grocery stores and is terribly high in mercury. Independent random tests have found the dolphin meat to contain 3 to 3,500 times the levels deemed safe by the Japanese Government.

Louis and Charles had shown me footage they had shot here in Taiji a year earlier: Thirteen small boats hunt the dolphins. Each boat has a long pole extended into the water, and when the fishermen find a pod, they bang on the poles, which confuses and scares the animals. The boats line up so that they form a kind of acoustical net and the men drive the pod into the narrow cove and seal it off. The next morning at dawn, they herd the dolphins—mothers, babies, whole families—by twos and threes into the shallows. Divers in black wet-suits and old round facemasks grab individual dolphins and drag them to the small gravel beach. Nooses leashed to a main line hang into the water. The divers, aided by fishermen in waders, wrestle the animals' tails into the nooses. The panicked dolphins blow and squirm. A man wades over to one and plunges a two-foot-long t-handled spike into her just behind her blowhole. There is a gush of blood, and the dol-phin thrashes and shivers. She screams, tips over onto her side and into the water, and, unable to clear her blowhole, begins to drown. All down the line more men are doing the same to others in the pod. The cove fills with blood. Terrified babies thrash around, driven crazy by the cries of their mothers. The dolphins tied to the beach quiver and writhe. If one of the fishermen thinks of it, he will wade over and cut a throat or spike one again. Many take half an hour to die.

Sometimes, when they've corralled thirty or forty dolphins or pilot whales—another species of large dolphin—the whalers will dis-pense with the niceties and simply drive their open boats through the pod, spearing and spiking. The wounded and dying tip over and cap-size in the blood, and some of the dolphins that are still swimming will try to get under them with their noses and hold them up.

OPS was making a feature film about the killing called *The Rising.* They had a lot of help from Ric O'Barry, the dolphin trainer for the old TV show *Flipper,* who had devoted his life to getting international attention to stop the slaughter—so far without much luck. Working closely with dolphins, he had never ceased to be amazed by their uncanny intelligence, emotions, and communication skills. "I knew dolphins were self-aware thirty years before the studies confirmed it," he told me. "On the show there was a dock. At one end of it was the house where the family was supposed to live, and at the other end was Flipper. They didn't actually live there, but I did. Every Friday night I'd drag a long extension cord and TV set out to the end of the dock and Flipper would watch *Flipper*. She loved it."

In some ways, O'Barry felt responsible for the cruel harvest. He believed that it was his show that engendered an international love affair with dolphins. He believed that what actually drove the killing in coves like Taiji was the capture of animals that were culled out of the slaughtered pods and sold to marine parks and swim-with-dolphin programs around the world. A prime female could net the fishermen $10,000, and the broker who then sold her to the parks could get $150,000. "If it weren't for the money involved in selling live dolphins," he told me, "this industry would have rightfully died years ago."

He had been sneaking into Taiji for the last five years to observe the killing and had become so well known to the locals he often had to drive through town in a wig and dress. He had shown Louis and Charles some video he had surreptitiously shot in the cove, and from the moment they saw it, they knew they had a subject for a movie. Louis also had a personal connection to the mercury issue: In the course of years of traveling and photographing the world's coasts and oceans, he had eaten tons of fish, and his blood levels for mercury were critically high. And Charles, as a world-class sailor, had loved the oceans and its dolphins all his life. All three men believed that here in Taiji, and in other coastal communities that consumed cetaceans, another Minimata scandal was in the making—that thousands of people would be sickened, crippled, or killed by mercury poisoning. But the greater issue of their movie was the alarming levels of mercury in *all* fish. Go on the Internet, they told me, look up FDA advisories for tuna fish, sea bass, and swordfish. You will be shocked. If people knew the danger, OPS believes, they would stop eating seafood, period. And there is an even greater issue beyond toxic fish: What put most of the mercury and PCPs into the oceans was the burning of fossil fuel, the same engine that drives global warming. So their message is that we are all responsible for this poisoning of the sea.

I did not want to see the killing in person. Even in the pitch dark, the inlet below our perch felt like a black hole, like concentrated dread. In a few hours, killing or no, activists from around the world would drive up to the public beach at the head of one finger of the inlet and paddle surfboards out into the mouth of the killing cove. They would form a surfer's circle, or paddle-out, which is the traditional ceremony surfers hold for a departed comrade. Flowers are tossed into the center and the surfers hold hands and send up a prayer for their friend's last big ride.

This surfer's circle for the dolphins was organized and led by Dave Rastovich, a beloved Australian surfer who had founded an organization and website called surfersforcetaceans.com. A few years ago, a dolphin had charged out of nowhere and saved him from an attacking shark, and since then he has worked hard to try to protect cetaceans. He had sailed on the *Farley Mowat* with Sea Shepherd in the Galapagos, and after some long conversations with Watson he had become galvanized in his activism. He figured that the only way to get the world to pay attention was to bring celebrities here to protest, and film it. He had invited Hayden Panettieri, the young star of the TV show *Heroes,* and Isabel Lucas, an Australian star. The two had traveled with a whale conservationist to San Ignacio Lagoon in Baja California and had observed breeding and calving gray whales at very close quarters for days. Both women said they had experienced a profound connection with the whales and that they would never be the same again. Karina Petroni, a top U.S. surfer, would be there, as well as James Pribram, the pro surfer who has launched an eco-warrior project, saving surf breaks around the world from pollution and overdevelopment.

Since Japanese culture does not respect or understand banner-waving protest, Rastovich thought it would be more in tune with a ritualistic, tradition-bound society to perform this simple ceremony honoring departed souls. He considers dolphins to be the original and best surfers, and anyone who surfs with any regularity will at some point end up on a wave with a dolphin and know that it's true. Rastovich was quite sure that this would be the first time in history that a paddle-out was enacted for nonhuman surfers. If I could get down off the mountain in time, I would join them.

By first light, the rain had stopped. We could see just beyond our hands, and we covered the cameras with leafy camo coats, dabbed on face paint, and sat very still. The cove took shape around us. It was hollowed out of spires of rock; gnarled trees spilled down their slopes. Mist tangled in the branches and shrouded the sharp spurs. Far below, waves crashed against the cliffs, and guard rocks fleeced with spume ran up to the shore like fleets of longboats. It looked like a scene from the classical Japanese pen-and-inks I had loved all my life. The cove was a natural fortress, and I could see the two fingers: the one running away to our right that cupped the public beach, and the inlet just across from us, hidden from all other views but ours, where the killing took place. The sky lightened and the water flushed a muted jade green. Two large gold hawks drifted in lazy circles over our blind. It was one of the most beautiful coasts I had ever seen.

I was relieved to see that no dolphins swam against the nets. On the narrow beach lay the rolls of green tarps they used to shroud the inlet. There would be no killing today.

We had the paddle-out ceremony. Thirty of us paddled surf-boards into the cove and tossed flowers and sent up prayers. Rastovich spoke some words about the thousands of dolphins that had died here, wishing them peace. It sounds a little hokey, but everyone had seen clips of the OPS footage, and as we sat on our boards just at the mouth of the killing cove it was difficult not to look in there and overlay pictures of violence.

The cops came, of course, and took everybody's passport info, and some angry fishermen pulled up in their trucks and fumed and stomped on their cigarettes. But all in all it went without a hitch. That night at the hotel in Osaka everybody felt pretty righteous and relieved—no violence, no arrests, no blood in the water—and the participants had a party.

The next afternoon Ric O'Barry called from Taiji, where he keeps watch on the cove, and said a pod of thirty pilot whales had been driven into the inlet and netted in, and would be killed the next morning. I guess the whalers thought the international attention had come and gone.

Six of us went back. We drove most of the night in a crowded van. Just out of Taiji, in the first faint light, we pulled over and got into wetsuits. Now there would be blood. The fishermen would have spears and knives, and they were known to get worked up and violent. When Allison and Alex had jumped into the cove with the dolphins several years earlier, they were hauled out of the water by the throat, thrown in jail for weeks, and fined. The OPS teams had gone back in earlier in the night—avoiding patrols, covering razor wire with rolls of black carpet—and rigged a battery of cameras. They stood watch over the cove, hidden again in the trees.

Taiji is all about whaling—they have a whale museum and an aqua park behind it, where dolphins captured from the cove perform tricks and tourists eat dolphin sandwiches. They have smiling dolphin statues and pictures everywhere. At dawn we parked behind a life-size bronze of a humpback with calf and waited for Charles, who called five minutes later: "They're killing! Go!" We sped to the public beach where we had been two mornings before. The six of us grabbed our boards from the van and splashed into the water, and we began to paddle as hard as we could toward the mouth of the killing cove. The plan was to perform another surfers' circle before we were stopped by the fishermen or the police. O'Barry wanted the footage to air around the world.

We scraped over one net line, and then it was forty yards of open water to the cove mouth. No boat appeared. We paddled past the rock corner and looked into the inlet. It was all blood. Thick and red, like paint. The dark bodies of two pilot whales floated in it, washed against the beach. And then I saw them: Twelve or fifteen of the little whales pressed in panic against a far net stretched across the cove.

They huddled tighter as we got closer. They circled in terror, blowing hard. We stopped, circled, and held hands. The blowing of the whales slowed. I watched the whales, twenty yards beyond us, slow and mill. They spy hopped, lifting their heads out to look. Small babies nosed out of the water to peer at us. They calmed down. Half of their group had just been butchered in this water, many crying for a long time before they died. These whales were in shock. But they seemed to sense that we meant them no harm. Rastovich wanted badly to cut the nets, but we could see another net stretched across the cove eighty yards out, and another beyond that. All these whales would be spiked within the hour.

We watched the remainder of this pod of pilot whales begin to flow against the net, their backs silvering in the long sun. We could hear them blowing. I sat on my board and felt tears stream down my cold face. This is what editors call going native, getting too close to a story. That was the farthest thing from my mind.

A boat cruised around an outer point of rock—a long, open motorboat. The fisherman throttled when he saw us. He skimmed over the lines of nets and wheeled dangerously close, yelling. He motioned for us to go. I looked around our circle. Everyone seemed calm. The whaler revved his motor and tried to frighten us with the propeller. He came so close to Hayden's leg where she sat on her board that she had to pull it out of the way. Furious, he circled once more and headed to the beach.

We paddled closer to the whales. I was sure the police must be on the way. In Japan you can be held for twenty-eight days without being charged, and no one was looking forward to a free cell in Taiji. We were just a few yards away from the pilot whales now. The little pod shrank against the net and we could hear them breathe fast, hollow blows over the slick water. We floated in the blood of their family. Hayden began to cry quietly. Then Isabel. Then Rastovich's wife, Hannah.

Soon the boat sped out to us again, and this time there were four whalers aboard. They feinted with the prop. One yelled wildly, picked up a long forked pole, and jabbed it at the closest boards. He hit Hannah in the thigh, then shoved Hayden's board. Both women stayed calm, keeping their balance, and holding the circle. Behind them I noticed the pilot whales going crazy, thrashing against the net. Enough. Rastovich said, "Let's paddle in. Stay against the rocks."

We ran from the water, threw the boards in the back of the van, ducked to the floor, and sped away. Forty-five seconds later, sirens wailed and police cars flew past us, heading toward the cove.

Back in Osaka we all changed flights and left the country before the authorities decided to make someone pay.

At home, I was amazed at the press the protest was getting. *People,* CNN, AP, MSNBC, *Ellen.* Hayden was everywhere, a real-life hero. Ric O'Barry, who had stood by the cove heartbroken and alone for so many years, was thrilled.

I wrote my story. When I tried to sleep, the image returned: those twelve pilot whales swimming against the nets, watching us. I watched them back. There is a lot I want to communicate. A baby lifts its head. They are wondering why we are there, but they don't feel threatened. A boat approaches. I hear the motor like a snarl. The whales circle tighter, faster. I lie beside my sleeping wife and weep. Months later, I continue to be shaken by a grief whose power I can't explain.

Perhaps it is that the little group of whales spoke, to me at least, for the whole ocean. The oceans are dying. Like the pod in the cove, half of what was once vital in the oceans is gone. Half the coral reefs. 90 percent of the great predator fish. What remains is in great peril. This year, along the California coast, the Chinook salmon simply did not show up in anything remotely like the usual numbers. The gray whales passing by, heading south for their breeding grounds in three lagoons in Baja, were shockingly emaciated. Whale observers and scientists wondered why.

I do not wonder why. These losses are happening at a faster and faster rate around the world. We are at the tipping point. People in the future will ask us what we did to keep the oceans from dying. *What did you do?*

April 8, 2008
Zihuatanejo, Mexico

Afterword to the
Tenth Anniversary Edition

In the ten years since the publication of *The Whale Warriors,* a lot has happened to Paul Watson, to Sea Shepherd, to the oceans.

I met Paul recently on a cool morning in September. He was speaking that night at a Boulder, Colorado, high school auditorium and attending a scuba convention, where he was mobbed for autographs. We stepped into a dive shop under the sandstone slabs of the Boulder Flatirons—about as far inland as a person can be. The captain had the same silver mane and beard, the heavy shoulders, the dogged look that lit quickly with mischief and laughter. But he seemed whiter, shinier, as if the intervening decade of salt and waves and sun had bleached and polished him like a piece of driftwood.

If he was notorious when I first met him, he was legendary now. The cable channel Animal Planet had taken care of that. And Sea Shepherd had become an eco-navy since our Antarctica campaign in 2006. I had loved that trip with its force 8 gales, forty-foot seas, and cunning naval tactics that allowed a patched-together former trawler to attack and ram whaling ships with ten times the tonnage and twice the speed.

The year after my voyage on the *Farley Mowat,* Watson had procured a Scottish fisheries patrol vessel that could keep up with the whaling fleet. He named it the *Robert Hunter,* and the next season the new attack ship, along with the *Farley,* rammed and drove the whale spotter vessel *Kaikō Maru* into the ice, saving several pods of minke whales. Then, in late 2007, Animal Planet began production of their show *Whale Wars.* The producers put cameras and crew on board Sea Shepherd's ships and just let Watson be Watson. This was the media-savvy activist who said, "If a camera doesn't record it, it didn't happen." The result, of course, was spectacular. The videographers closely followed the eco-pirates as they harassed, obstructed, rammed, and documented Japan's illegal whaling fleet. It fast became the most-watched show the channel had ever had, attracting more than a million viewers for the first season's finale and lasting an astonishing seven years. The mild-mannered but steely Peter Hammarstedt, commanding his ship on the bridge, became a television celebrity. Chris Aultman also became a star—the intrepid helicopter pilot who once choppered us eighty miles along a desolate Antarctic ice edge that broke and pieced itself together in a frigid mosaic. I remembered how exposed we felt on that flight—to sudden weather or mechanical failure or getting lost—and how I thought that if we had trouble finding the little *Farley Mowat* again in all that ocean and burned fuel looking for it, we might die.

A generation of viewers was transfixed by the Shepherds, who were willing to die to save a whale. I think more than the maritime combat, what attracted an audience most was the imagination and esprit de corps of the eco-pirates. They attacked the whaling fleet in zodiacs and patrol boats, with water hoses and stink bombs and prop foulers, with the hulls of the ships that kept them afloat. And though they witnessed minke whales with new calves being harpooned and exploded, and saw the babies swimming terrified in the blood of their mothers, they still kept their sense of fun and humor. Sea Shepherd was swamped with applications. New chapters sprang up in countries around the world. Money and attention from the show, and continued support from wealthy donors, allowed them to buy more ships, such as the combat-ready flagship *Bob Barker—* which Hammarstedt commanded and which looks in its maritime camouflage like an eco-destroyer—and to finally have an onboard hangar for Chris's helicopter so it wasn't constantly doused and buried with corrosive salt water.

Sea Shepherd greatly expanded its mission. Watson said that they now are conducting more than three dozen campaigns around the world at the same time. They increased patrols in the Galápagos Marine Reserve against poachers. In 2009 they gave Ecuador a sophisticated AIS marine tracking system, which allows the government to follow every boat in the national park. Sea Shepherd launched a campaign in cooperation with the Mexican government to protect the critically endangered vaquita porpoise, the smallest cetacean in the world, in the Sea of Cortez. Peter Hammarstedt now commands an operation in West Africa, working in conjunction with the governments of Gabon and Liberia, to fight illegal fishing in their waters. There are campaigns to stop poaching in Sicily, to destroy ghost nets in the Mediterranean, and to save sea turtles in Costa Rica and the Mozambique Channel. My friend the artist Geert Vons became director of Sea Shepherd Netherlands, and Alex Cornelison is now the CEO of Sea Shepherd Global. Watson told me that Hammarstedt caught the *Thunder*, a notorious poaching ship, fishing in the Southern Ocean off Antarctica and engaged in the longest pursuit of a poacher in history. "Peter chased him for a hundred and ten days, up the west coast of Africa to the waters off São Tomé, until the *Thunder* was running out of fuel and they sank their own ship in front of him. They were trying to destroy the evidence. But our crew boarded the ship as it sank and documented the evidence. The captain was sentenced to three years in prison and fined 15 million euros."

"We're working with many governments," Watson said. "We can do what they can't. We now have twelve ships. We're operating in forty-five countries. And right now there's 250 volunteers on those ships." He said they've gotten more sophisticated in using the courts and have launched a legal arm that is pursuing cases against illegal fishing and whaling around the world.

But what he seemed most proud of was the ability of Sea Shepherd to inspire individuals to action. He said, "Anyone can change the world. A graphic designer in Holland called me up. He said they were slaughtering seals on one of the Orkney Islands, what was I going to do about it. I said, 'What are *you* going to do about it?' He got a group together, and they confronted the sealers and threw their rifles into the ocean. They filmed it, and the media attention helped raise money, and we bought the island, which is now a seal sanctuary."

Watson himself is landlocked for now. In 2012, when he landed in Frankfurt on his way to Cannes, Costa Rica exercised a warrant against him through Interpol for an incident that had occurred ten years earlier. Watson says the prime minister of Japan had just met with the president of Costa Rica and requested the action. German border officers detained him, and he was put under house arrest. Two months into it, "I got tipped off. Somebody . . . called up my lawyer and said on Monday . . . I would be deported to Japan." He said that he slipped out to Holland and Sea Shepherd France picked him up in a sailboat from Amsterdam and whisked him across the Atlantic to Nova Scotia. "They had taken my passport. I entered the States with a California drivers license." He was picked up at Catalina Island, and "we crossed to American Samoa where I met the *Steve Irwin* in time for the [Antarctica] campaign. I went almost around the world with no ID."

The wily sea fox had escaped the clutches of the law again. But as of this writing he can't leave the United States because he is on Interpol's Red Notice, which he says, with maybe just a touch of pride, "is for serial killers, major drug traffickers, and war criminals. I'm the only person in history to be put on there for conspiracy to board a whaling ship."

He was becoming animated, the same energized and ever-mischievous Watson who once astounded an incredulous port captain in Melbourne who had just seen the all-black pirate ship *Farley Mowat* come into view. Port control saw the attack ship flying its Jolly Roger and said over the radio, "*That* is a pleasure craft?" And Watson shot back, "We enforce international conservation law. I take great pleasure in it." Now he said, "You know, it's not about me. We don't have one leader anymore, we have dozens of leaders—the directors of the various countries. We're decentralized. We have become a movement. All over the world. Bureaucrats and governments can try to destroy me. You can stop an individual—what the Japanese tried to do to me—and you can stop an organization, but you can't stop a movement."

Watson admits that what Sea Shepherd is trying to do—save the oceans, whales, sharks, sea cucumbers, reefs, seabirds, and tiny plankton—is getting harder. The odds grow longer. In 2014 Australia took Japan to the International Court of Justice, arguing that their whaling program was illegal, and won. Japan desisted for a year and then in 2016 declared an exception to the ruling and began to whale again off Antarctica. The Japanese government passed an "antiterrorism" law making it a terrorist act to harass their ships, or even to document the cetacean slaughter in Taiji. And the whaling fleet received help from the Japanese government in the form of military satellite technology that allowed them to track Sea Shepherd ship positions and movements in real time. Watson said, "We knew where they were, but we could no longer close the gap." Unable any longer to physically reach the whalers, this year Sea Shepherd suspended its antiwhaling campaign in the Southern Ocean. He said, "Overall, we saved 6500 whales directly. Possibly over 11,000 because 40 percent of the whales they kill are pregnant. We exposed the whole thing to the world through *Whale Wars*, and the new quota is 333 whales instead of 1035. And so every year we're saving 702 whales. I think we accomplished all we could possibly do within the confines of practicality and the law."

Other challenges loom. He told me global warming is the greatest threat to the planet and that governments aren't willing to do anything about it.

Then he got an expression I remembered from the bridge of the *Farley Mowat*. Something compounded of empathy for all the living things that cannot speak for themselves, the birds and whales, the corals and turtles that are getting wiped off the planet; and in the expression, too, was a ferocious defiance. He said, "We don't do what we do because we've figured out the odds, or that we are going to win or lose. We do it because it's the right thing to do."

September 2017
Denver, Colorado

Acknowledgments

I am greatly indebted to my first readers, Kim Yan and Leslie Heller. And also to Lisa Jones and Helen Thorpe for their energy, astute suggestions, and valued friendship. Thanks, too, to Susan Swarmbrunn for her support, and to Gillian Silverman for her generosity and scholarship. Brian Barker was a great help.

Special thanks to my Texas Hold 'em playing agent, David Halpern, and to the rest of the Robbins Office, for their tireless work on my behalf.

To two scientists and esteemed friends, astrobiologist David Grinspoon and paleobotanist Kirk Johnson: thanks for not only the deep well of knowledge, but also for the refreshing wind of the Long View.

Thanks to Cliff Ransom, Stephen Byers, John Rasmus and *National Geographic Adventure* for launching the assignment to Antarctica. I am especially grateful to Brad Wieners for championing the story. And to Laura Hohnhold for her historical support.

To Elizabeth Stein, my editor—a bow of the deepest gratitude.

Index

Peter Heller is the author of the novels *The Dog Stars* and *Celine* and the surfing memoir *Kook*. He is an award-winning adventure writer and longtime contributor to NPR. He is a contributing editor at *Outside* magazine and *National Geographic Adventure*. He lives in Denver, Colorado.